William Thayer Smith

Elementary Physiology and Hygiene

The Human Body and its Health

William Thayer Smith

Elementary Physiology and Hygiene
The Human Body and its Health

ISBN/EAN: 9783337371579

Printed in Europe, USA, Canada, Australia, Japan

Cover: Foto ©berggeist007 / pixelio.de

More available books at **www.hansebooks.com**

Elementary Physiology and Hygiene

THE HUMAN BODY

AND

ITS HEALTH

A TEXT-BOOK FOR SCHOOLS, HAVING SPECIAL REFERENCE TO THE EFFECTS OF STIMULANTS AND NARCOTICS ON THE HUMAN SYSTEM

BY

WILLIAM THAYER SMITH, M.D.

ASSOCIATE PROFESSOR OF ANATOMY AND PHYSIOLOGY IN DARTMOUTH MEDICAL COLLEGE

IVISON, BLAKEMAN, TAYLOR, & COMPANY

PUBLISHERS

NEW YORK AND CHICAGO

EXTRACTS FROM RECENT LEGISLATION

AFFECTING THE PUBLIC SCHOOLS.

STATE OF MINNESOTA.

SECTION 1. That all school-officers in the State may introduce as part of the daily exercises of each school in their jurisdiction, instruction in the elements of social and moral science, including . . . self-denial, health, purity, temperance, cleanliness. . . .

STATE OF MICHIGAN.

SECT. 15. The district board shall specify the studies to be pursued in the schools of the district: *Provided always*, That provision shall be made for instructing all pupils in every school in physiology and hygiene with special reference to the effects of alcoholic drinks, stimulants, and narcotics generally, upon the human system.

STATE OF NEW HAMPSHIRE.

. . . And in physiology and hygiene, with special reference to the effects of alcoholic drinks, stimulants, and narcotics upon the human system.

STATE OF NEW YORK.

Provision shall be made by the proper local school authorities for instructing all pupils in all schools supported by public money, or under State control, in physiology and hygiene, with special reference to the effects of alcoholic drinks, stimulants, and narcotics upon the human system.

STATE OF OHIO.

To the board of education the State has primarily intrusted the responsibility of seeing that her teachers are grounded in all the dangers that beset her youth, and that they are disposed to arm their youth with the requisite knowledge, convictions, and resolution to guard against these dangers. The effect of alcohol on the brain and nerves—its innate tendency, like all narcotic poisons, to enslave the appetite, and lead to excess and ruin—should be clearly made known to every youth while in school. —*From notes explaining school-law.*

STATE OF VERMONT.

. . . And elementary physiology and hygiene, which shall give special prominence to the effects of stimulants and narcotics upon the human system. . . .

*** Several other States have passed similar laws.

PREFACE.

In making this little book, I have tried to give to the student a definite impression, in outline, of the structure and functions of the human body. To this end I have tried to omit all statements that would confuse the picture by overloading it, and all statements that could not be understood by those who will be its most numerous readers. I have not told them, for example, that the re-action of the saliva is alkaline, and that of the gastric juice acid, because for many of them that statement would have no meaning. I have not mentioned the names of many of the muscles, because it is difficult and unnecessary to remember them.

The laws of hygiene are given in connection with the facts of anatomy and physiology from which they are derived. Learned in this way, they will remain in the mind as guiding *principles,* and not simply as the dicta of authority.

In treating of the effects of stimulants and narcotics, I have endeavored to set forth facts which are susceptible of abundant proof, and which are of the most importance, practically, to those for whom this work is designed.

WILLIAM THAYER SMITH.

HANOVER, N.H., August, 1884.

CONTENTS.

CHAPTER I.

CHAPTER II.

CHAPTER III.

CHAPTER IV.

CHAPTER V.

CHAPTER VI.

CHAPTER VII.

CHAPTER VIII.

CHAPTER IX.

CHAPTER IX. — *Concluded.*

CHAPTER X.

APPENDIX.

LIST OF ILLUSTRATIONS.

COLORED ILLUSTRATIONS.

THE HUMAN BODY.

CHAPTER I.

DEFINITIONS.

Section I.—1. If we wish to study a machine, such as a clock or a steam-engine, we take it to pieces, and examine each part separately. We inquire what each part is called, what it is made of, and how it fits in with the other parts. We then ask what is the use of each part, and how it works. Knowing these things, we understand the machine and its action.

2. The human body, which is the most wonderful of all machines, is to be studied in this way.

Anatomy names and describes its parts. It tells us their size, weight, shape, color, texture, and composition; their position, and relation to other parts.

Physiology acquaints us with the action of each part, and the work that it does. It tells us how it acts, when it acts, what makes it act, and what is the effect of its action.

Anatomy may be studied in the lifeless body.

Physiology must be studied in the living body.

Anatomy is well known, because all the parts of the body have been carefully studied and described.

Physiology is only partially known. There are some parts of the body whose use we do not know. In those parts with which we are better acquainted we find much that we do not understand. But many learned men are devoting their whole time to this study, and are constantly adding to our knowledge.

By combining the teachings of Physiology with what we know by experience we construct the science of **Hygiene.** From its principles we derive rules for the preservation of health.

Anatomy is a science of *Structure.*
Physiology is a science of *Function.*
Hygiene is the science of *Health.*

QUESTIONS.

SECTION I.—What is the natural method of studying the body? What does Anatomy tell us of the body? What does Physiology tell us of the body? Can Anatomy be studied in the living body? Can Physiology be studied in the lifeless body? Which is most thoroughly known,—Anatomy or Physiology? Define Anatomy. Physiology. Hygiene.

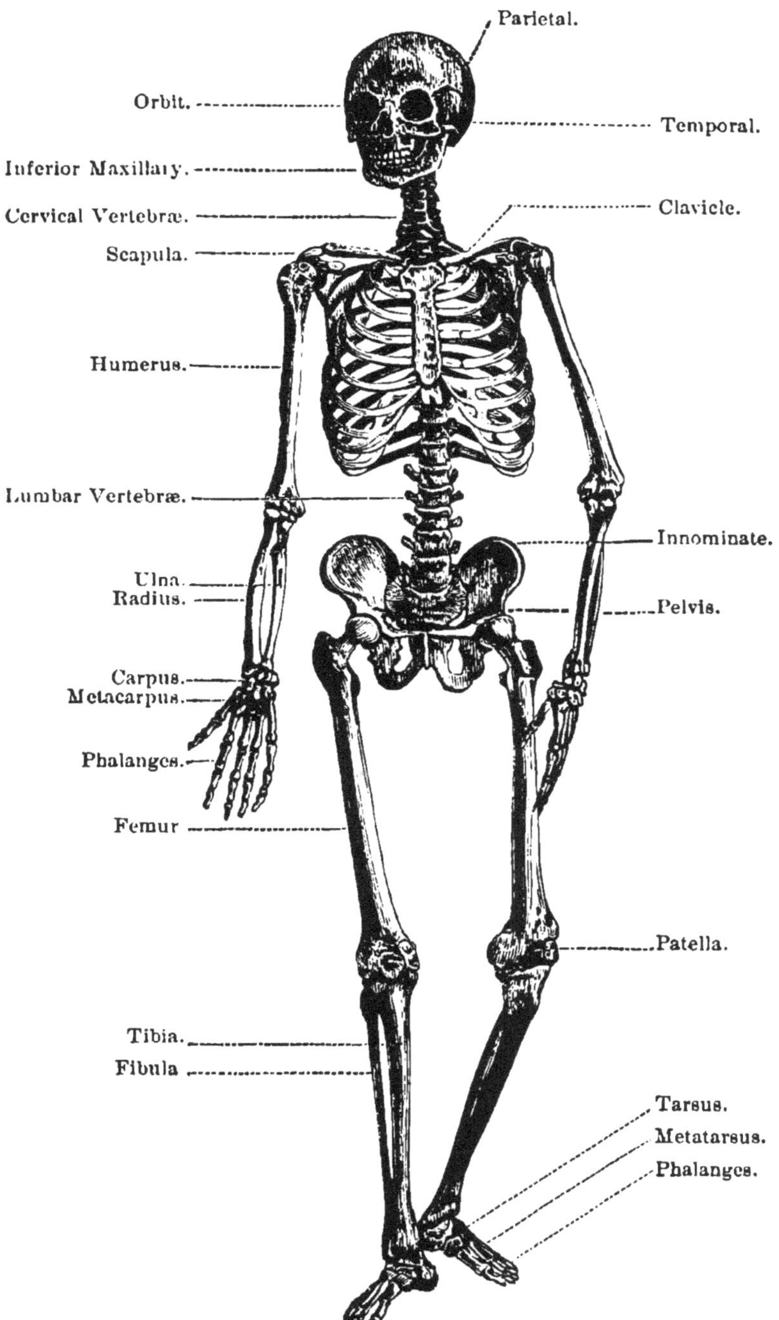
Parietal.
Orbit.
Temporal.
Inferior Maxillary.
Cervical Vertebræ.
Clavicle.
Scapula.
Humerus.
Lumbar Vertebræ.
Innominate.
Ulna.
Radius.
Pelvis.
Carpus.
Metacarpus.
Phalanges.
Femur
Patella.
Tibia.
Fibula
Tarsus.
Metatarsus.
Phalanges.

Fig. 1.

THE SKELETON.

The Head, 28 Bones.

Occipital (*base of skull*)	1	**Palate** (*back part of roof of mouth*)	2
Parietal (*sides of skull*)	2	**Lachrymal** (*in orbit*)	2
Temporal (*temples*)	2	**Malar** (*cheek-bones*)	2
Frontal (*forehead*)	1	**Superior Maxillary** (*upper jaw*)	2
Sphenoid (*behind the face*)	1	**Inferior Maxillary** (*lower jaw*)	1
Ethmoid (*behind the face*)	1	**Malleus** (*in the ear*) very small.	2
Nasal (*bridge of nose*)	2	**Incus** (*in the ear*) very small.	2
Vomer (*between nasal fossæ*)	1	**Stapes** (*in the ear*) very small.	2
Turbinated (*on walls of nasal fossæ*)	2		

The Trunk, 52 Bones.

Vertebræ (*backbone*)	24	**Hyoid** (*in the neck*)	1
Sacrum (*backbone*)	1	**Ribs**	24
Coccyx (*backbone*)	1	**Sternum** (*breast-bone*)	1

The Upper Limbs, 64 Bones.

Scapula (*shoulder-blade*)	2	**Ulna** (*fore-arm*)	2
Clavicle (*collar-bone*)	2	**Carpus** (*wrist*)	16
Humerus (*arm-bone*)	2	**Metacarpus** (*hand*)	10
Radius (*fore-arm*)	2	**Phalanges** (*fingers*)	28

The Lower Limbs, 62 Bones.

Innominate (*hip-bone*)	2	**Fibula** (*leg*)	2
Femur (*thigh-bone*)	2	**Tarsus** (*ankle, heel, instep*)	14
Patella (*knee-pan*)	2	**Metatarsus** (*flat of foot*)	10
Tibia (*leg*)	2	**Phalanges** (*toes*)	28

CHAPTER II.

THE BONES AND JOINTS.

SECTION I.—1. The bones are the framework of the body. When joined, as in the living man, they constitute the skeleton. They serve three purposes:—

1. They give the body shape and firmness of outline. The soft parts which cover them add grace.

2. They act as levers by which the muscles attached to them move the body.

3. They protect important organs.

2. There are two main cavities in the body formed wholly or in part by the skeleton; viz.,—

1. **The cavity of the skull and spinal column.** The skull contains the brain, and is a tight box whose walls are strong, and which is so shaped as to resist great pressure. There are no openings into it except those small ones through which blood-vessels and nerves pass in and

---

SUGGESTIONS TO TEACHERS.—1. Every school in which anatomy and physiology are taught ought, if possible, to have a human skeleton. Lacking this, the next best thing is to have the skeleton of some quadruped. The general resemblance will be sufficient to make it a good illustration of the text. If you have no complete skeleton, get dried bones,—beef-bones, mutton-bones, vertebræ, long bones, jaw-bones. Have them sawed in different directions. Much can be learned from them. The differences between dried bone and fresh, living bone must, however, be borne in mind. The experiments of softening a bone by maceration in weak hydrochloric acid for a few weeks, and of removing the animal matter by burning, are easily tried.

2. The different parts of a joint can be shown in a sheep's leg.

out. The great opening at the base of the skull connects its cavity with the spinal canal, and is fully an inch in diameter. The other openings are very much smaller than this.

The spinal canal is also well guarded by its walls, and by bony projections. It contains the spinal cord.

2. **The cavity of the trunk.** This is divided by a horizontal partition, called the diaphragm, into two parts. The upper part is the thorax, or chest; and the lower, the abdomen and pelvis.

The **thorax** is a bony cage formed by the backbone behind, the ribs at the side, and the breast-bone in front. It contains the heart and lungs. These organs need to be guarded against blows or pressure, but they are not so easily injured in this way as the brain and spinal cord. The thorax is, therefore, not so close a box as the skull. Moreover, it is needful that the walls of the thorax should be movable in order that we may breathe. The thorax is so made that it gives sufficient protection to the organs which it contains, and at the same time it can enlarge and contract.

The **abdomen** is not as well guarded as the thorax. There is no bony wall in front; and the intestines are easily wounded, though they bear pressure and displacement much better than the heart and lungs. But if the abdomen were walled in front, like the thorax, we could not bend our bodies. For purposes of motion, and to permit the expansion of the intestines after a full meal, they are left partially unprotected.

The **pelvis** (Latin, *pelvis,* a basin) is formed by the hip-bones and the sacrum and coccyx. Its contents are well guarded by those thick bones.

3. There are 206 bones in the body:—

In the head			28
In the trunk	spinal column	26	52
	ribs	24	
	sternum	1	
	hyoid	1	
In the upper limbs			64
In the lower limbs			62
			206

BONES OF THE HEAD AND TRUNK.

SECTION II.—1. The **skull** is poised on the top of the spinal column, and contains twenty-eight bones. In the young infant they are loosely united, but in time many of them become welded together so that they can not be separated. The lower jaw, and the small bones of the ear, are the only ones of them that are movable. Besides containing the brain, the skull protects the organs of hearing, of smell, of vision, and of taste.

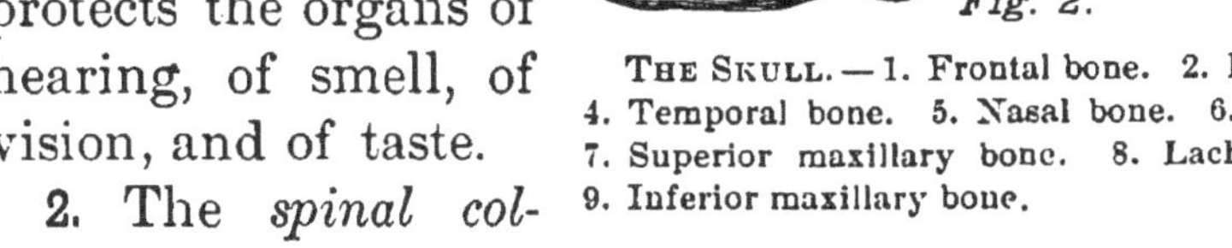

Fig. 2.

THE SKULL.—1. Frontal bone. 2. Parietal bone. 4. Temporal bone. 5. Nasal bone. 6. Malar bone. 7. Superior maxillary bone. 8. Lachrymal bone. 9. Inferior maxillary bone.

2. The *spinal column*, or *backbone*, consists of twenty-four vertebræ, the sacrum, and the coccyx. The **vertebræ** (Latin, *vertere*, to turn) are so called because they form an axis on which the body turns. They are irregular in shape, and consist of

a body and an arch. They have projections called *processes,* to which muscles are attached. When they are joined in position, the arches together constitute the spinal canal, and the bodies form a solid column for support. Between the bodies are cushions of a tough and elastic substance called fibro-cartilage. Each is about one-fourth of an inch thick, and is firmly united to the bone above and below. They serve as springs in the column, and allow a twisting motion.

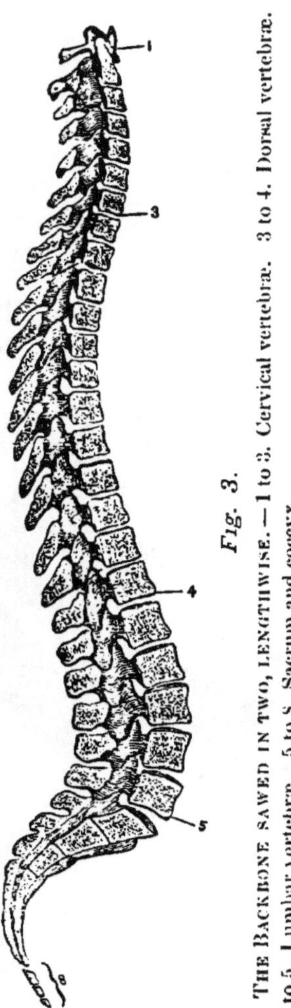

Fig. 3.

THE BACKBONE SAWED IN TWO, LENGTHWISE. — 1 to 3. Cervical vertebræ. 3 to 4. Dorsal vertebræ. 4 to 5. Lumbar vertebræ. 5 to 8. Sacrum and coccyx.

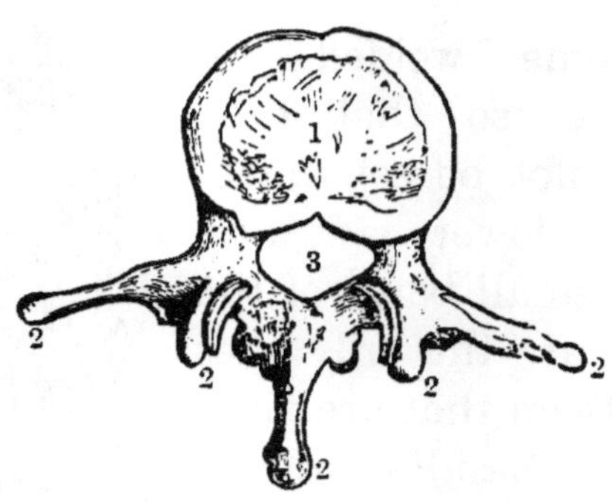

Fig. 4.

A VERTEBRA. — 1. Body. 2. Processes. 3. Spinal canal.

The **sacrum** is a wedge-shaped bone, which fits in between the hip-bones.

The **coccyx** is the end of the column, and is a small, curved bone, commonly in two or more pieces, which are united by joints.

3. The spine in a baby is perfectly straight, and his back flat. As he grows and walks the spine becomes slightly curved backward in the region of the shoulder-blades, and forward at the waist. This is natural. But frequently the curve of the back becomes too great. The shoulders are drawn forward, and the chest flattened. This makes a stoop. It is caused by weakness of the muscles which sustain the back and head, or indolence and carelessness. It makes an ungraceful shape, and is injurious because it compresses the heart and lungs, and checks their free action. An erect posture and a full chest should be cultivated. To this end, the muscles of the chest and trunk must be kept vigorous by exercise.

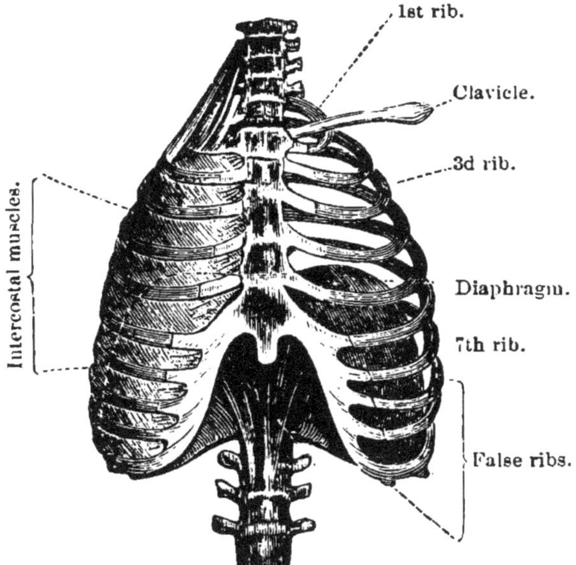

Fig. 5. THE THORAX.

The natural curves of the spine are forward and backward. A curve to one side is unnatural and a deformity. It comes from bodily weakness, lack of air and exercise, and from standing on one foot habitually, or sitting much of the time at a desk or table with one arm resting upon it while the other hangs down.

4. The ribs and breast-bone with the backbone form the thorax. The ribs are twenty-four in number, twelve on each side. They are joined to the backbone behind. The first seven are called true ribs. They are connected with the breast-bone by the costal cartilages. The last five are called false ribs. They are not directly connected with the breast-bone. The last two false ribs are called floating-ribs, because their front ends are not joined to any bone.

The **costal cartilages** are continuations of the ribs. Cartilage is more elastic than bone, and the wall of the thorax expands and contracts more freely for being partly cartilage. It is by this expansion and contraction that we breathe. Whoever has harnessed horses has noticed that they generally swell their chests, and manifest displeasure, when the girths are buckled tight. It interferes with their breathing, and so with their comfort. If we are wise we shall resist, as they do, any compression of our chests or waists by tight clothing. The elastic walls readily yield to pressure, and after a time become permanently misshapen. The heart and lungs are then crowded; and the liver is displaced, and encroaches on the other organs. Thus, a figure is acquired which is neither beautiful nor healthful.

Fig. 6.
THE STERNUM.

5. The **sternum**, or breast-bone, is a flat, narrow bone, about five inches long, which is in the middle line of the chest in front. The collar-bones and the cartilages of the ribs are joined to it.

6. The **hyoid bone** is a slender bone shaped like a horseshoe. It is situated in the neck just above Adam's apple, where it may easily be felt. The base of the tongue is attached to it.

BONES OF THE UPPER LIMB.

SECTION III.—1. The upper limb is divided by anatomists into *shoulder*, *arm*, *fore-arm*, and *hand*. In ordinary language, we call all between the shoulder and the hand, the arm. In anatomical language, the arm extends from shoulder to elbow; the fore-arm, from elbow to hand.

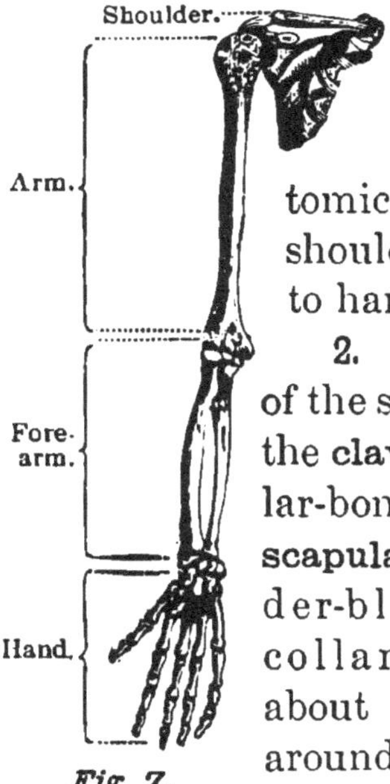

Fig. 7.
THE UPPER LIMB.

2. The bones of the shoulder are the **clavicle**, or collar-bone, and the **scapula**, or shoulder-blade. The collar-bone is about as large around as a finger, and curved in shape. It extends from the upper corner of the breast-bone to the shoulder-blade. It braces the shoulder.

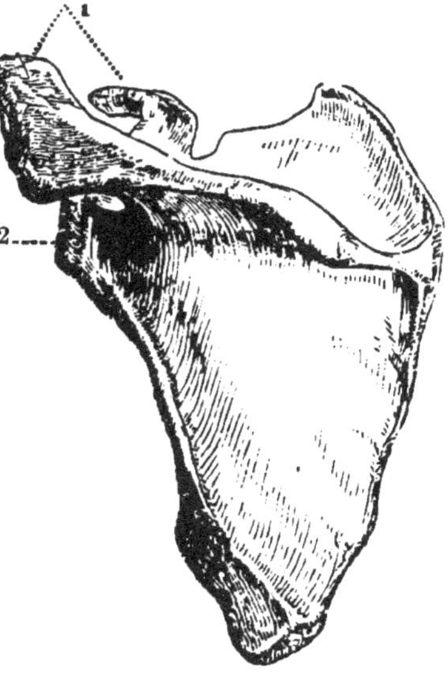

Fig. 8.
THE SCAPULA.—1. Processes. 2. Surface for shoulder joint.

3. The shoulder-blade is a flat, three-cornered bone, with projections extending from it, to which muscles are attached. At one corner is a smooth

surface slightly hollowed out, on which the head of the arm-bone plays.

4. The arm-bone is called the **humerus.** It is a strong bone, about a foot long in a grown man. It has a round head, which plays on the joint surface in the shoulder-blade. Its lower end is joined to the two bones of the fore-arm.

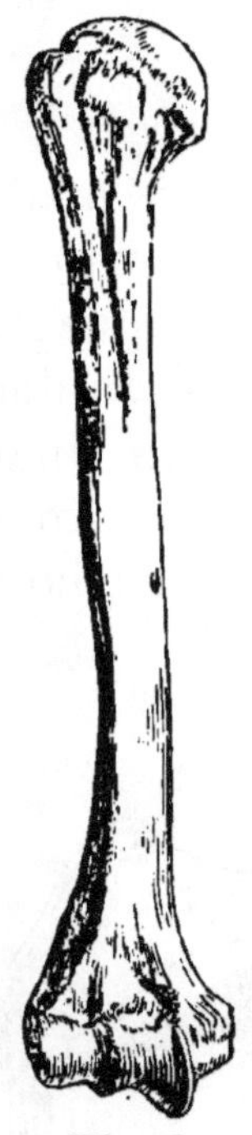

Fig. 9.
THE HUMERUS.

5. The bones of the fore-arm are called the **radius** and the **ulna.** The radius is on the thumb side, and the ulna on the side of the little finger. The radius is joined to the humerus and to the ulna in such a way that it rolls over on the ulna, and turns the palm of the hand up or down. The ulna is so joined to the humerus that it can only move forward and backward. The radius is joined closely to the hand, and, when it rolls over the ulna, carries the hand with it.

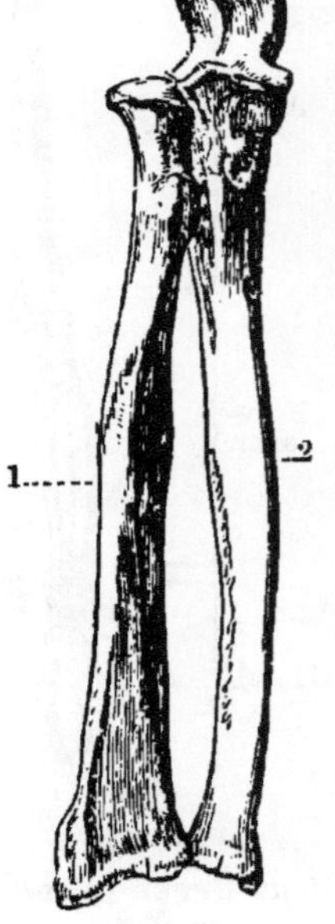

Fig. 10.
1. THE RADIUS.
2. THE ULNA.

6. The hand is divided into the **carpus,** or wrist, **metacarpus,** or palm, and **phalanges,** or fingers.

7. The wrist has eight short bones, irregular in size and shape. These are bound by ligaments into a compact bunch. They glide a little on each other. Though they seem to be put together without design, they are really shaped and joined in such a way as to give freedom of movement combined with strength.

8. The five metacarpal bones are slightly curved, so as to make a hollow in the palm.

The phalanges are in three rows, and are so called because they are like rows of soldiers (Greek, *phalanx*, a body of soldiers).

The thumb stands out from the rest, and can be made to meet the end of each of the fingers. This enables us to pick up and handle small things with great delicacy. None of the lower animals has a thumb like man's except a few of the apes, and theirs is not so perfect for handling. It is his hand more than any other part, except the brain, that gives man his superiority over them. Its skill and delicacy when trained are wonderful. Most of the work of the world is done in part with the hand. Very much of it could not be done at all if man's hand were not as perfect as it is.

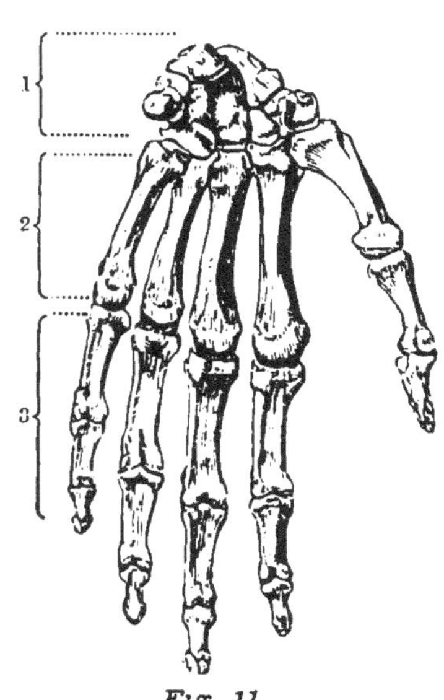

Fig. 11.

THE HAND.—1. Carpus. 2. Metacarpus. 3. Phalanges.

BONES OF THE LOWER LIMB.

SECTION IV.—1. The lower limb is divided into *hip*, *thigh*, *leg*, and *foot*. In ordinary language, the word leg means the lower limb from hip to foot. In anatomical language, the part between the hip and knee is called the thigh; the part between the knee and foot is the leg.

2. The hip-bone is so irregular in shape, that the old anatomists could not think of any name that suited it; and so they called it the *os innominatum*, nameless bone.

The two hip-bones come together in front. Behind they are separated by the sacrum. The cavity inclosed between them is the **pelvis,** or basin.

3. The **femur,** or thigh-bone, is the longest bone in the body. It has a round head, which fits into a socket in the hip-bone. At its lower end it spreads out to make a broad surface for the knee-joint. It is slightly curved.

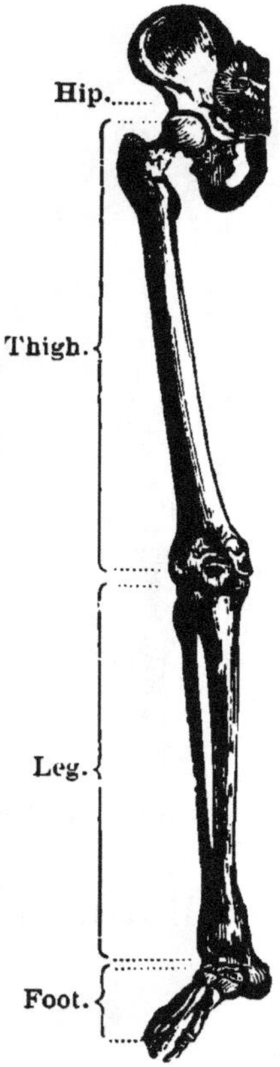

Fig. 12.
THE LOWER LIMB.

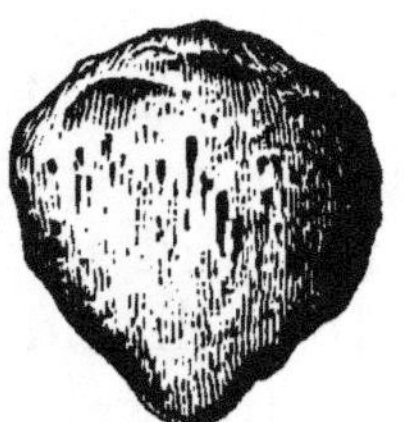

Fig. 14.
THE PATELLA.

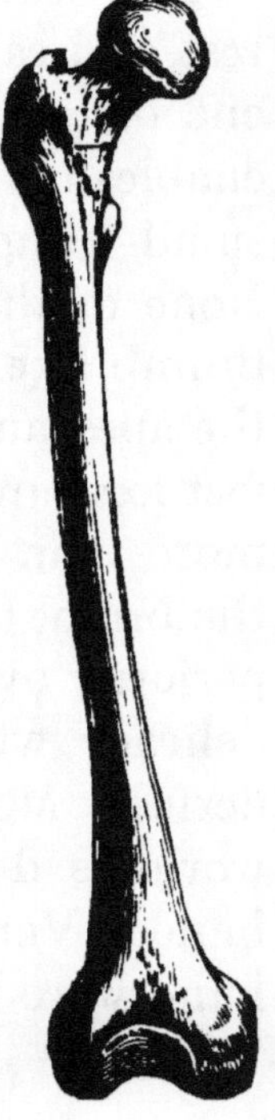

Fig. 13.
THE FEMUR.

4. In front of the knee is a small bone, heart-shaped, which is called the **patella,** or knee-pan.

5. The leg has two bones, the **tibia** (Latin, *tibia,* a flute) and the **fibula** (Latin, *fibula,* a shawl-pin). The tibia is a strong bone. It has a sharp ridge in front, which is called the shin. The tibia is joined to the femur above. The fibula is a long, slender bone. It is joined to the tibia above and below. The lower ends of these two bones are joined to the ankle-

bone of the foot. They can easily be felt, one on the inner, the other on the outer, side of the ankle.

6. The foot consists of **tarsus, metatarsus,** and **phalanges.**

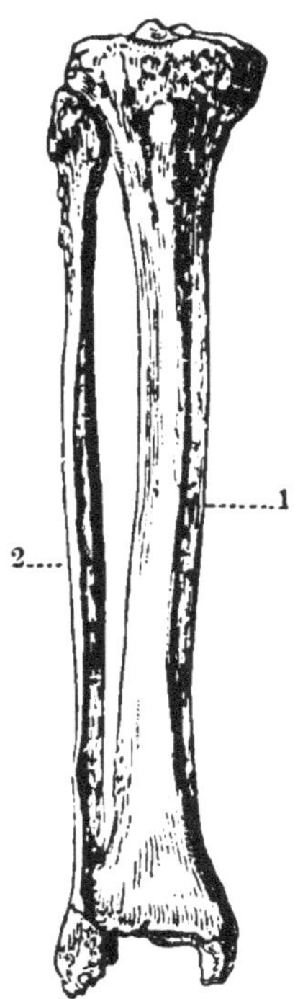

Fig. 15.

1. THE TIBIA.
2. THE FIBULA.

7. There are seven irregular bones in the tarsus. They form the ankle, the heel, and the instep.

The metatarsal bones form the "flat" of the foot, and part of the instep. There are five in each foot.

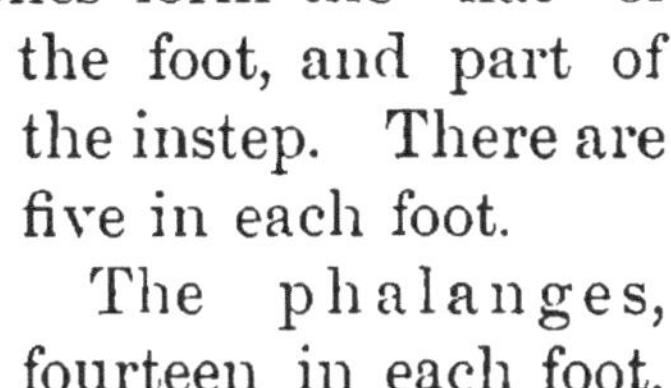

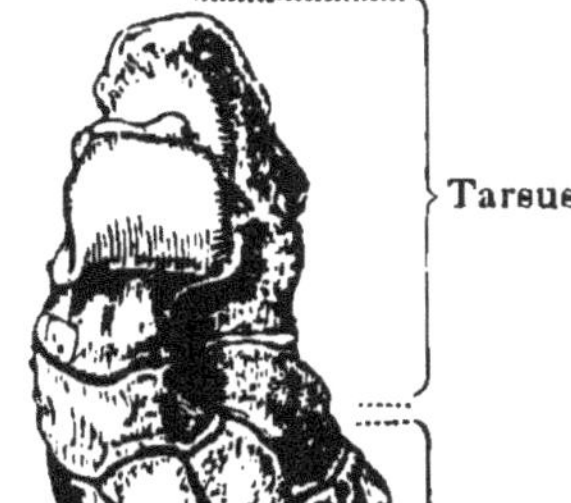

Fig. 16.

BONES OF THE FOOT.

The phalanges, fourteen in each foot, form the toes.

8. The bony structure of the foot is quite like that of the hand. The differences are such as fit it for its humbler work. Its inner side is arched, and the weight rests on the heel and the ball of the toe. The foot is strong and elastic, and should be dressed in such a way as not to distort its shape or check its movement. But, while we wonder at and despise the Chinese practice in this regard, we treat our feet in ways as truly unreasonable. The shape of our shoes is determined, not by the design of Nature, but by fashion. They are often too tight. Almost always they are too short, and too nar-

row across the toes. Consequently the toes of most grown people are squeezed together, frequently overlapping. The great toe naturally rests a little separated from its neighbor, and almost in a straight line with the inner side of the foot. If we make it turn in, its principal joint stands out prominent, and from the pressure becomes inflamed. A bunion is formed there. A corn is another painful result of pressure.

The foot is a much-enduring member, and is useful, even though crippled. But grace and ease of movement are constantly sacrificed, and numberless miseries incurred, for the sake of making the foot look small.

Athletes and pedestrians long ago discovered the necessity of a shoe which gives freedom to the foot. Such a shoe is never tight. The heel (if any) is low and broad, and directly under the heel of the foot. The sole is as broad as the foot itself, and at least half an inch longer.

STRUCTURE OF BONE.

SECTION V.—1. If we saw a long bone in two lengthwise two things are noticeable:—

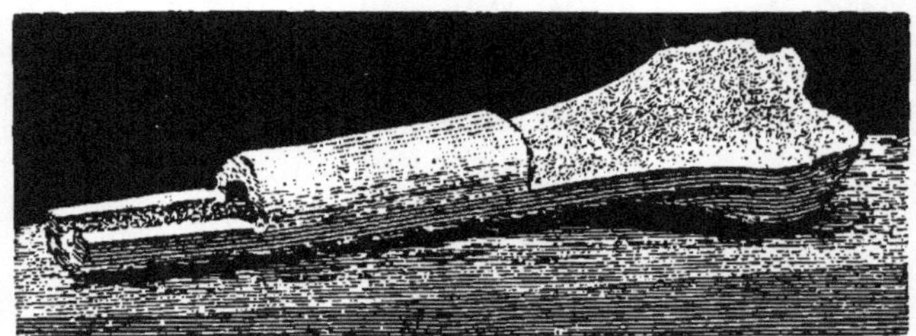

Fig. 17.
SECTION OF THE FEMUR.

1. Its shaft is hollow. During life the cavity is filled with marrow. This consists largely of fat, and is a store of nourishment which helps to sustain the body when it is deprived of food. It has been proved by experiment, that a hollow shaft is stronger than a solid shaft of the same material and the *same weight* and length.

2. The bony substance of the shaft is hard and compact. The large ends, on the contrary, though they are not hollow, are filled with cells, and present a honeycombed appearance. It is desirable that the shaft should be slender and strong. Its substance is, therefore, very compact. It is desirable that the ends should be large, to give a broad surface for the joints. Their substance is, therefore, open. If it were compact, it would increase the weight unnecessarily.

2. If we saw a flat bone in two, as one of the bones of the skull, or a short bone as one of the bones of the wrist, we shall find, that, while its shell is of compact tissue, its inside is of the same honeycombed tissue that we found in the ends of the long bone.

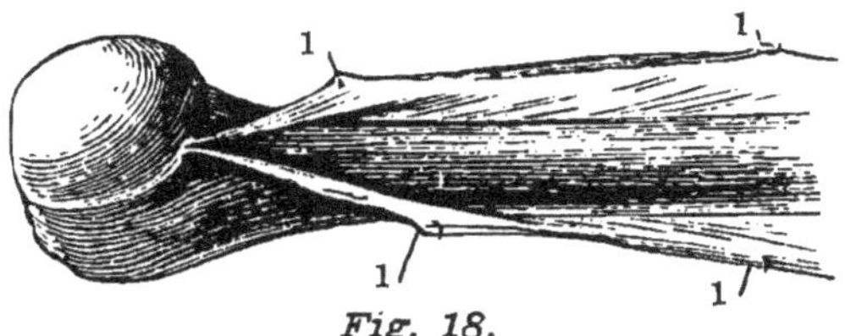

Fig. 18.
1. THE PERIOSTEUM.

3. Bones are well supplied with blood-vessels and nerves. We can find in every bone one or more holes through which a small blood-vessel passes to the interior. A living bone is covered with a membrane called the **periosteum.** In this membrane is a fine net-work of blood-vessels, from which a countless number of little vessels pass directly into canals in the bone. These are called Haversian canals. They run through every part of the bone.

CHEMICAL COMPOSITION OF BONE.

SECTION VI.—1. Bone is about two-thirds mineral, and one-third animal, matter. The mineral matter is chiefly phosphate of lime. This substance makes more than half of the bone. The mineral and animal matter are closely combined, but they can be separated in two ways:—

1. If we put a bone in a moderate fire, and burn it, it will become porous and brittle, but will retain its shape. The animal matter is burned out, but the mineral remains.

2. If we put a bone in weak hydrochloric acid, and allow it to remain for a few weeks, it will become as pliable as a rope, and can be tied in a knot. The acid has eaten out all the mineral matter, but has left the animal matter. If a bone has too much mineral matter, it is brittle. This is the condition of the bones of old people, and sometimes of younger persons. They break very easily. In the bones of children, on the other hand, the animal matter is abundant; and they will bend a good deal before they break. In the disease of children called rickets, there is so little mineral matter in the bones, that they are too soft. The legs become bowed, the head enlarges, and the whole frame gets out of shape.

Fig. 19.
A BONE AFTER SOAKING IN HYDROCHLORIC ACID.

2. During childhood and youth the skeleton is assum-

ing the form which it is to keep through life. It is pliable, and may be molded to healthfulness and grace, or to deformity. Nature should be allowed to shape it in her own way, and all habits of dress or attitude or movement that interfere with the natural outlines should be carefully avoided.

THE JOINTS.

SECTION VII.—1. The joints of the body with which we are most familiar are movable; but most of the joints of the head-bones are like a joint in a table or chair. They are fixed.

The joints between the bodies of the vertebræ have a little motion, but are not freely movable like the joints of the limbs.

There are, then, in the body,—

Immovable joints.
Slightly movable joints.
Freely movable joints.

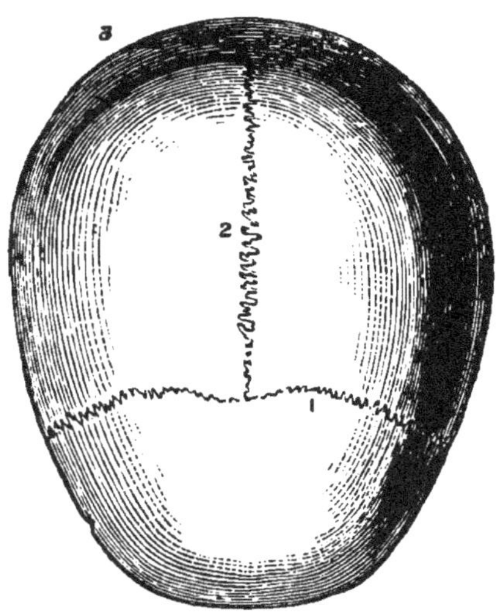

Fig. 20.

JOINTS OF THE SKULL.—1. Frontal bone. 2. Parietal bone. 3. Occipital bone.

2. The movable joints are of various kinds. Sometimes the two surfaces only glide a little on each other, as in the wrist. Sometimes one bone moves on another, like a door on its hinge. The elbow and the ankle are of this kind.

Sometimes a round bone fits into a round socket, and moves in all directions. The hip-joint is of this kind.

PARTS WHICH COMPOSE A JOINT.

3. In a joint there are—

1. Two or more bones, each covered, where they come in contact, with a thin, smooth layer of cartilage or gristle. Cartilage is more elastic than bone, and serves as a spring in the joint.

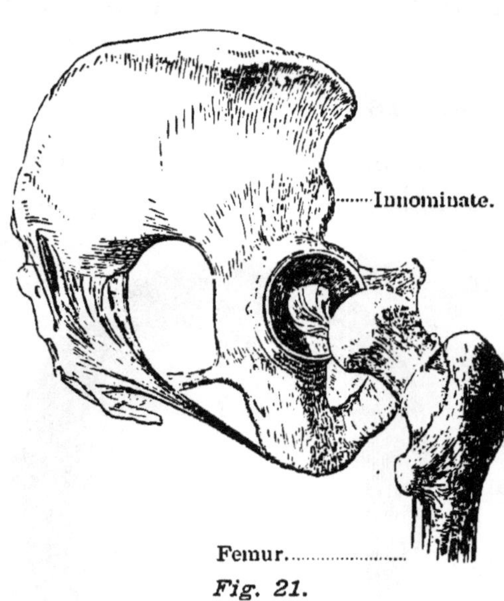

Fig. 21.
THE HIP-JOINT.

2. Ligaments which bind the ends of the bones firmly together. A ligament is a white, glistening band, very strong, and generally not elastic. When we examine it with a microscope, we find that it is made of fine white fibers, lying side by side like the threads of a ribbon, only there are no cross-fibers. They all run lengthwise.

3. The cavity of the joint has a thin lining, called a *synovial membrane.* This membrane gives out a fluid called synovia, or joint-water. This serves the same purpose that the oil we put in the joints of machinery serves. But while these need constant attention, and soon wear out, the living joint oils itself, and may be in constant use for seventy years or more without causing a thought in the mind of its owner.

If, however, joints become diseased, they are very

painful. They sometimes become enlarged and stiff and misshapen.

4. The joints of young children bend very freely: as they grow older, they become less flexible. Some have "looser" joints than others. Those public performers who can twist themselves into marvelous shapes, are persons, who, by a course of training begun early in life, have gradually stretched their ligaments.

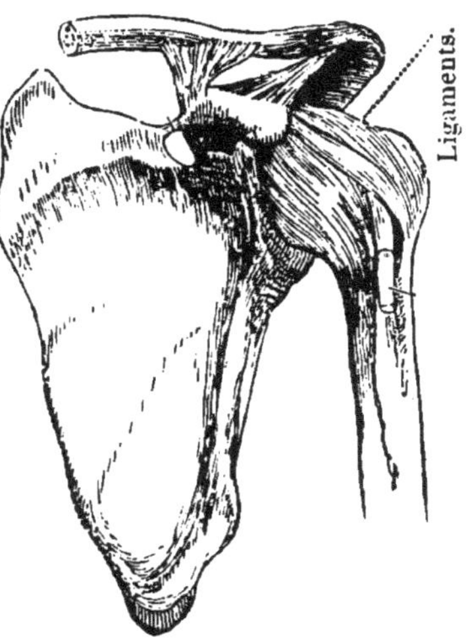

Fig. 22.
THE SHOULDER-JOINT.

5. Ligaments are tough, and not easily hurt; but when we do "sprain" a joint, which means that we have torn or overstretched its ligaments, they recover slowly.

QUESTIONS.

SECTION I.—1. What is the skeleton? What three purposes does it serve? Can an animal body destitute of bones move? If so, in what way?

2. What are the two main cavities formed by the skeleton? How are the skull and backbone especially fitted to protect their contents? What are the two main divisions of the cavity of the trunk? What is that portion of the lower division which is between the hip-bones called? What is the name of the partition between chest and abdomen? Is there any partition between abdomen and pelvis? What are the bony walls of the chest? What are its principal contents? Why is not the abdomen as well walled as the thorax?

3. How many bones are there in the body? In the spinal column?

In the head? In the trunk? In the upper limbs? In the lower limbs?

SECTION II.—1. What bones of the skull are movable? What organs of special sense are guarded by the bones of the skull?

2. What is the structure of a vertebra? What is a *process* on a bone? What canal do the vertebræ make when they are joined? With what cavity is this canal continuous? What springs are found in the spinal column? What is the position of the sacrum? Of the coccyx?

3. What is the shape of a baby's spine? What are the natural curves of the back? What curves are unnatural? What is the harm of a stoop? What are the causes of unnatural curves in the backbone?

4. How are the first seven ribs distinguished from the last five? Which are called floating-ribs? What are the costal cartilages? What is their use? What are the effects of compressing the chest and waist?

5. Describe the sternum.

6. Where is the hyoid bone situated? and what important organ is attached to it?

SECTION III.—1. What are the divisions of the upper limb?

2. What is the common name of the clavicle? What is its situation?

3. What is the common name of the scapula? What is its situation?

4. Name the arm-bone.

5. What are the bones of the fore-arm? Which rolls over the other? Which is most closely connected with the hand?

6. What are the divisions of the hand?

7. How many bones in the carpus?

8. How many bones in the metacarpus? How many phalanges? What in the human hand especially gives it an advantage over the hands and paws of lower animals?

SECTION IV.—1. What are the divisions of the lower limbs?

2. Why is the hip-bone called the os innominatum?

3. What is the longest bone in the body?

4. Where is the patella situated?

5. Name the leg-bones. What is the shin? What is the use of the fibula?

6. What are the divisions of the foot?

7. How many bones in the tarsus? In the metatarsus? How many phalanges?

8. What is the effect of wearing too tight shoes? Of having a high heel under the middle of the foot? What is the natural direction of the great toe? What is the proper shape for a shoe?

SECTION V.—1. Is a bone solid all through? What part is hollow? What part is filled with thin-walled cells? What is contained in the cavity? Why is it hollow? Why are the ends of the long bones enlarged?

2. What is the structure of a short bone?

3. How do blood-vessels and nerves get into a bone? What are the Haversian canals?

SECTION VI.—1. What is the chemical composition of bone? How can the animal and mineral constituents be separated? Why do the bones of old people break more easily than children's?

SECTION VII.—1. What three kinds of joints are there in the body?

2. Of the movable joints, what varieties are there?

3. What are the essential parts of a joint? How does a joint oil itself?

4. Why are some people's joints looser than those of other people?

5. What is a sprain?

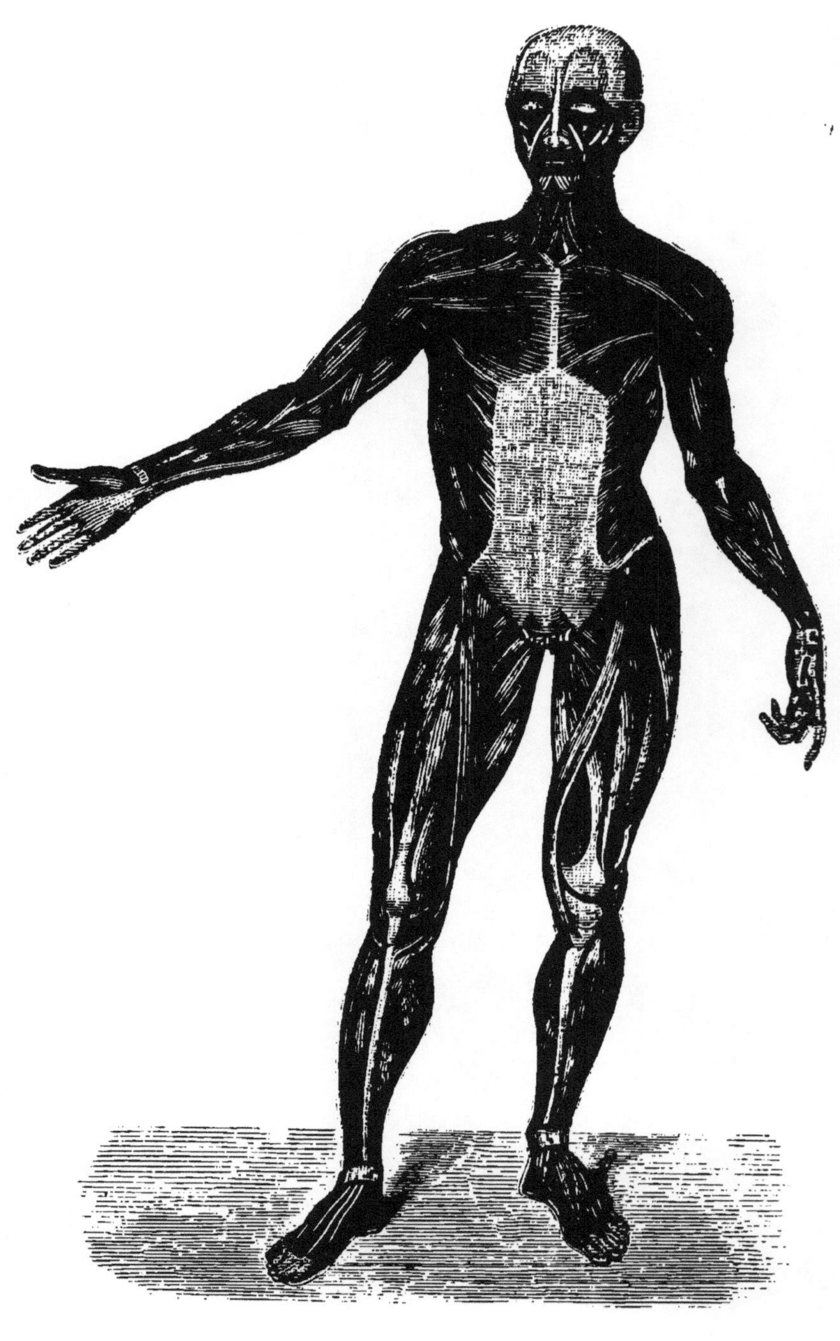

THE FIGURE ON THE OPPOSITE PAGE GIVES A GENERAL REPRESENTATION OF THE MUSCULAR SYSTEM.

CHAPTER III.

THE MUSCLES.

SECTION I.—**1.** The muscles are organs of motion. They cover the bones, and constitute the "flesh." They are found also in the walls of the alimentary canal, and of the blood-vessels, and in other inward parts. The heart is chiefly muscle.

2. The muscles with which we are most familiar, act when we *will* that they shall act. We can walk, run, or sit still; we can move our arms or our heads as we will; but we can not stop the beating of our hearts in this way. The movements of the intestines are also independent of our will.

3. Muscles are accordingly divided into two classes,—**voluntary,** or those which are subject to the will; and **involuntary,** or those which are not subject to the will.

4. The voluntary muscles constitute about two-fifths of the weight of the body. They are compactly arranged over the skeleton, most of them being attached to a bone

---

SUGGESTIONS TO TEACHERS.—1. A piece of fresh beef will illustrate to the naked eye the gross structure of voluntary muscle. With a microscope magnifying four hundred times, a small fragment teased out with a needle will show the structure of the fibers. The play of the muscles and tendons in the fore-arm is easily seen in a thin person. The action of involuntary muscle can be seen in the iris, whose muscular fibers contract under the influence of light, but are not subject to the will.

2. Fibrous tissue is seen between the muscular bundles of beef. Tendons show in the leg of a fowl.

at each end. They are of various shapes, according to their position and use. Some are long, some short: some are round, and some flat. In the living body, all the muscles of a limb are bound together and covered by fibrous tissue.

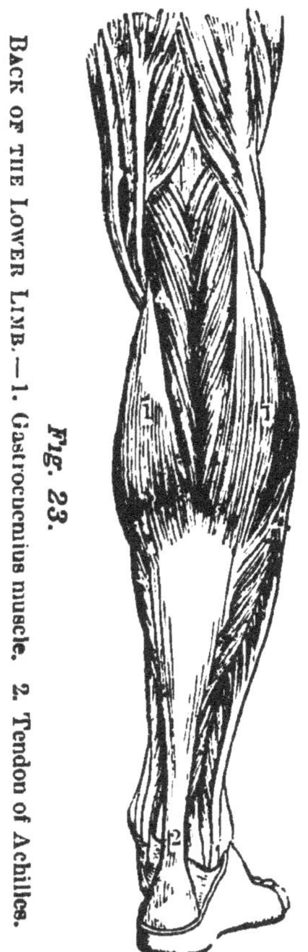

Fig. 23.
BACK OF THE LOWER LIMB.—1. Gastrocnemius muscle. 2. Tendon of Achilles.

5. Voluntary muscles are commonly attached to a bone, at one or both ends, by means of a **tendon** or **aponeurosis.** A tendon is a glistening cord of fibrous tissue. It is tough, and does not stretch. An aponeurosis differs from a tendon in being flat. If the muscle is flat, it ends in an aponeurosis: if it is round, it tapers down to a tendon.

6. There are more than five hundred muscles in the body. Most of them are in pairs, the two sides of the body being alike. The smallest is the **stapedius,** a muscle in the ear, which is only one-sixth of an inch in length: the longest is the **sartorius,** which extends from the hip to the leg below the knee, and is over eighteen inches in length.

The **biceps** of the arm has two heads, both arising from the shoulder-blade, and is inserted into the radius just below the elbow. The great muscles which pass from the breast-bone to the upper end of the arm are called **pectorals.** The **gastrocnemius** constitutes a large part of the calf of the leg. Its tendon, called the **tendon of Achilles,** is inserted into the heel, and is the largest tendon in the body.

SECTION II.—1. Muscle has a peculiar power of shortening itself. There is a kind of elastic tissue in the body, of which a few ligaments are made, which, like rubber, will contract after it has been stretched; but muscle is the only tissue that contracts without being stretched first. When a muscle grows shorter, it grows thicker at the same time, just as the body of a worm will shorten and thicken. If, for example, I place my left hand on the biceps of my right arm, and then bend my elbow, I shall feel the biceps swelling, and growing hard: at the

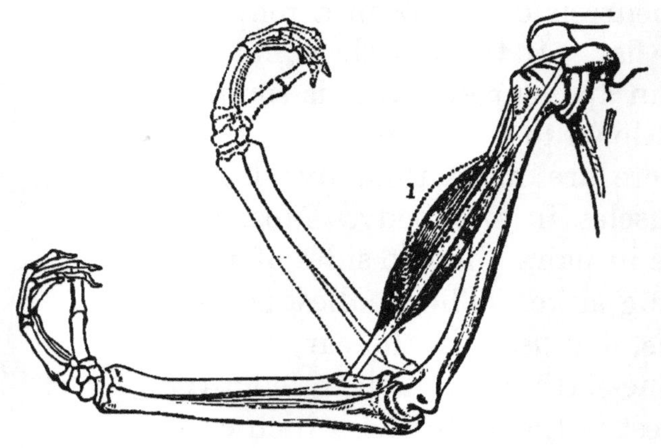

Fig. 24.

1. BICEPS MUSCLE. The dotted lines indicate the changed shape of the biceps when the fore-arm is drawn up.

same time it has shortened, and thus drawn the bone to which it is attached up toward the shoulder. We do not understand how it is that muscle contracts when we will that it shall. We can only say that it does so.

2. Involuntary muscle has the same power of shorten-

ing that voluntary muscle has. But it will not do this in obedience to the will. Cold contracts, and heat relaxes it. The involuntary muscle in the walls of the intestine contracts when food comes in contact with it. Many other influences produce the same effect, but the will which controls the voluntary muscle has no authority over the involuntary. Involuntary muscle never contracts as rapidly as voluntary muscle sometimes does. It moves in a sluggish way.

STRUCTURE OF MUSCLE.

SECTION III.—1. If we examine a little shred of voluntary muscle with a microscope which will magnify four hundred times, we shall see that the small fibers which are visible to the naked eye are made up of still smaller fibers, and that these have dark lines crossing them. These microscopic fibers are gathered into bundles, and these again into larger bundles. The muscle is made up of many such bundles, surrounded and bound together by fibrous membranes. We can readily see this in a piece of beef. The fibers constitute what we call the grain of the meat.

Fig. 25. FIBERS OF VOLUNTARY MUSCLE.

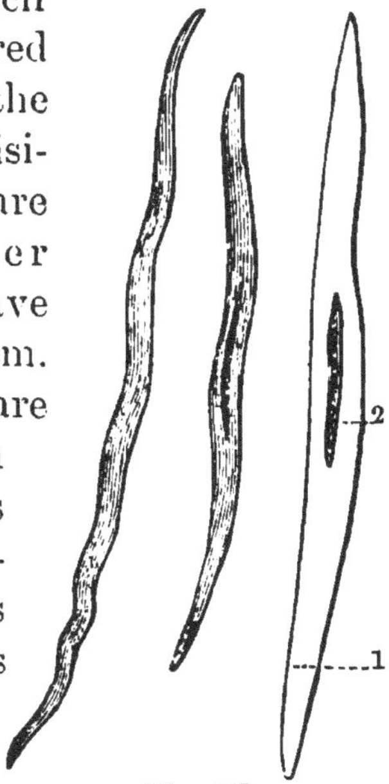

Fig. 26. FIBERS OF INVOLUNTARY MUSCLE.—1. Fiber soaked in acetic acid. 2. Nucleus.

2. Involuntary muscle, when examined with a microscope, is also seen to consist of fibers. But they are much shorter than

the fibers of voluntary muscle; and they are broad in the middle, and taper to each end. They have no cross-stripes, and have a spot in the center called a nucleus.

MUSCULAR EXERCISE.

SECTION IV.—1. It is a law of Nature, that every living thing, in order to remain in health, must do the thing that it was made to do. A stone was made to lie inactive and motionless. Men, as well as the lower animals, were made for activity and motion. Every part of the body must do the work that it was made to do. The muscles must move, the stomach must digest, the nerves must be in use, the brain must think. The part which is allowed to remain inactive becomes unhealthy and weak. This is the great law of exercise.

2. This law applies especially to young persons. After the body has got its growth, and has become firm and strong, it can endure bad usage or neglect better than in the growing period.

3. Some people are engaged in occupations which give abundant exercise to all the organs of body and mind. Such occupations tend to long life. Statistics kept in the State of Massachusetts for thirty years, of the length of life in different occupations, showed that the farmers lived the longest.

4. Many employments require the use of the muscles more than of the brain. Persons engaged in such employments should, during their hours of recreation, exercise their minds.

5. Other occupations engage the mind while the body is inactive. Those who have such occupations should seek recreation in muscular exercise.

6. It is especially important that the muscles should be exercised, for several reasons.

1. Because they are so large a part of the body. Nearly one-half the weight of a man is muscle.

2. Because they are made capable of great activity. They are abundantly supplied with blood-vessels, and the current of blood should flow freely through them. When they are idle, its flow is sluggish.

3. Because, when the muscles are exercised, all other parts of the body are refreshed. We may exercise the brain vigorously while the muscles are quite idle, and the blood circulates no faster. But, when we exercise the muscles, the heart beats more strongly, and the blood flows more rapidly, not only through the muscles, but also through the skin, the liver, the stomach, the brain. The organs which remove waste matters from the blood are more active, and the whole body is purified. Exercise is Nature's stimulant.

7. Muscular exercise is valuable, not only because it promotes the general health, but also because it directly improves the muscular system. A strong and well-shaped body is to be desired. The ancient Greeks, who for their physical beauty and vigor, and their intellectual power, have been the admiration of all succeeding nations, made very much of the cultivation of the muscles. Their scholars and statesmen were proud if they could win prizes in the great athletic games. Much of the work of life is done by the muscles. It will be better done if they are strong and well trained.

8. Three things are to be sought in the training of the muscles,—*1. Strength, 2. Alertness, 3. Endurance.*

1. Strength. The muscles increase in size and power by

use. It is important that this increase should be uniform. Exercises should be chosen with reference to the development of arms, legs, and trunk. It is a common mistake, to regard strength as the only end of muscular exercise. A man is not required to do the work of a horse. Great strength does not always imply good health. Athletes are sometimes overtrained, so that their health is impaired while their muscles are large. Prize-fighters frequently die early.

2. *Alertness.* This is the power to obey quickly the commands of the will or the impulse of the senses. In many trades, and some professions, it is of the utmost value. Games which require keen watching and rapid movements develop it.

3. *Endurance.* This is the power of continuing to make efforts for a long time without tiring out. It is not always the largest muscle that has the most of this quality. This, with healthy action of other organs, constitutes what is called "staying power," and gives the victory in a long race. It is acquired by regular exercise.

9. Muscular exercises are valuable, also, because they commonly train the eye and the ear to quickness of perception.

10. The muscles are larger and firmer in men than in women. Nevertheless, the need of muscular exercise is just as real for one sex as for the other. It is as truly requisite for their health and proper growth. Neither unsuitable dress, nor false ideas of propriety, should be allowed to deprive them of it.

11. Exercise, to be most useful, should be *regular.* To take several hours of it to-day, and none to-morrow, is less beneficial than to take a moderate amount daily. It

should be taken, if possible, *out of doors.* Much of its benefit comes from breathing a great deal of pure air.

The more heartily the mind is interested and engaged, the more the benefit.

12. Violent exercise should not be taken directly after a full meal. At this time the stomach has work to do. If the blood is drawn away from it to the muscles, and all the strength is engaged in muscular efforts, the digestion will be checked. If such a practice be long continued, the stomach will be weakened.

13. In muscular exercise, as in every thing else, it is important to avoid excess. It is excess to exercise so long or severely as to be unfitted for other occupations. It is excess to exercise until a muscle or limb is painfully exhausted, and does not soon rest. It is excess to attempt through ambition exercises that are beyond the strength. It is excess to engage in severe exercises when exhausted by mental or other labor.

Excessive exercise sometimes results in bleeding from the lungs, sometimes in enlargement of the heart. That organ, being compelled to overwork, becomes too large; and this causes illness, and sometimes death.

14. There are many small muscles in the face, which are attached by one end to the bone, by the other to the skin. By acting together, they give a great variety of expression. The thoughts and feelings are indicated. Grief causes one set of muscles to contract, joy another. So naturally is the action of particular muscles associated with certain emotions, that it is difficult for most people to conceal their feelings from one who is steadfastly observing the countenance. The faces of the lower animals are not as well supplied with muscles, nor are they as soft

and movable as the human face. They express only a limited number of thoughts and feelings. The face of the lion expresses dignity; that of the tiger, cruelty; that of the ox, patience. The expressions which are most frequently on the face become after a time permanent: thus the character is written on the countenance. A sullen or bitter temper makes an unpleasing aspect. A genial and kindly disposition will in time impart its own beauty to the face.

EFFECT OF ALCOHOL ON THE MUSCLES.

SECTION V.—1. By the action of alcohol, muscle is sometimes changed, in part, to fat. It thus becomes flabby and feeble.

Alcohol affects the muscles indirectly, by affecting the digestion and the blood, and so spoiling their nourishment.

The athlete training for a prize, knows well, that, if he indulges freely in alcoholic drinks, he will surely fail to bring his muscles to a hard and vigorous condition. Total abstinence from alcohol and tobacco are important for his success.

QUESTIONS.

SECTION I.—1. What are muscles? Where are they found?

2, 3. What two kinds of muscles are there?

4. What holds the muscles together?

5. What is a tendon? An aponeurosis?

6. How many muscles in the body? Which is the longest? Which is the smallest? Where is the biceps humeri? Pectoral? Gastrocnemius? Tendon of Achilles?

SECTION II.—1 What is the special property of muscle? Can we explain the contraction of muscle?

2. What is the difference between the contraction of voluntary and involuntary muscle?

SECTION III.—1. How does voluntary muscle look under the microscope?

2. How does involuntary muscle look under the microscope?

SECTION IV.—1. What is the law of exercise?

2. Is exercise more important for young than for older persons?

3. What class of people have the best prospect of long life?

4, 5. What kind of exercise is best suited to a muscle-worker? What kind to a brain-worker?

6. What three general reasons are there for muscular exercise?

7. What special reasons?

8. What three qualities of muscle should be cultivated?

11. Name the conditions which make exercise useful.

12. Name the cautions to be observed in exercise.

13. What is excess in exercise? What are its results?

14. What is the use of the muscles of the face? How is character written on the countenance?

SECTION V.—1. What change can alcohol make in muscle? How does it affect the muscles indirectly? What is the practice of athletes in training with regard to alcohol and tobacco?

CHAPTER IV.

WORK AND WASTE.—THE BLOOD.

SECTION I.—1. The body, like not-living machines, tends to wear out by use. Even the enamel of the teeth, which is the hardest substance found in it, at length gives way.

2. The soft parts wear out very rapidly. In every movement of a muscle, in every action of the brain, some of their particles are worn away.

3. The body is always active. Even in sleep, the breathing muscles, and the heart, and the muscles of the digestive organs, are moving. The wear is, therefore, constant.

4. But the body differs from a not-living machine in being able to *repair itself*. In early life, it does more than this. It builds itself up, and grows. When it has got its growth, it still repairs all waste, and may increase in strength and endurance for many years. At length there comes a period when it is not able to repair so much as it once could; it can not, therefore, endure so much wear,—the old can not be so active as the young; finally it is unable even to supply the waste caused by the simplest actions, such as breathing and digesting; then life must cease.

SUGGESTION TO TEACHERS.—The least stain of blood, on a glass slide, under a microscope magnifying four hundred times, is sufficient to show the red globules. It is not always easy to find white ones.

5. Few people die of old age. Generally, before the time for such a death is reached, some disease attacks the enfeebled body, and overcomes it.

6. The process of repair is going on all the time: but, during our waking-hours, waste is greater than repair; hence, every one must *sleep* several hours in each twenty-four. During sleep, the waste is very small; and the process of repair restores to the system each night what it has lost during the day.

7. The need of sleep is indicated by a feeling of fatigue and drowsiness. These are the warnings that Nature gives us to stop for repairs. Without this feeling, we would not be willing to lie inactive for so much of each day. Since rest is very important, this feeling is made so strong that it is almost impossible to resist it. Exhausted men will fall asleep on horseback, or even walking.

8. But Nature gives her warning in good time. If it is necessary to continue working for a while after we begin to feel tired, we can generally do so without injury; but we must rest afterward. However tired we may be at night, if we are rested by our night's sleep we are safe; but if the wear of the day is not fully repaired, and we feel each morning more weary than on the morning before, we are in danger.

9. A very young infant sleeps most of the time. He requires less sleep as he grows older. Grown persons need from six to nine hours sleep every day. Indolent people often take too much. On the other hand, it is easy to injure the health, particularly in early life, by taking hours which belong to sleep for work or pleasure. This is especially injurious when the weary body and

mind are excited by stimulants, to make them forget the need of rest.

10. Sleep, though the most perfect, is not the only, way of resting. When we have been using the muscles, we may rest them while using the brain. After study, hearty muscular *exercise* is rest. Entire change of occupation is rest.

MATERIALS FOR REPAIR.

SECTION II.—1. The materials by which the waste of the system is repaired, are found in **air, food,** and **water.** Air must be taken in constantly, food and water at frequent intervals. We can live but a few minutes without air: without water we should perish in a few days. Men have lived without food for weeks, having air and water; but the body must at length be exhausted if these three things are not regularly supplied in sufficient quantity.

2. We take oxygen from the air by the breathing apparatus. We take food and water by the digestive apparatus. Both are taken into the blood, and carried through the body, to be used as needed in every part.

3. The blood not only carries the oxygen, food, and water to every part of the body: it also carries away from every part the *waste matter* which is formed there, and delivers it to the organs which discharge it from the body. It has been likened to an express-agent, who goes about the streets of a city, leaving a parcel here, taking one there, — his wagon always loaded, though its contents are constantly changing.

THE BLOOD.

SECTION III.—1. **Blood** is a red fluid, not transparent, having a salty taste. It is a little heavier and thicker than water. If we look at a drop of it through a microscope, we shall see in it many small round bodies, shaped like a coin with a thick rim. These are only $\frac{1}{3200}$ of an inch across, and are all of the same size. They are called **red-blood corpuscles,** or *globules.* They make up nearly half the blood.

Fig. 27.
RED CORPUSCLES OF HUMAN BLOOD (400 Diameters).

2. Every animal that has a backbone has red globules in its blood, but they differ in shape and size. In birds and reptiles they are not round, but oval, in shape, and have a spot in them called **a nucleus.** In all animals that have round globules, they are smaller in size than those of a man's blood, with two exceptions, the elephant and the sloth.

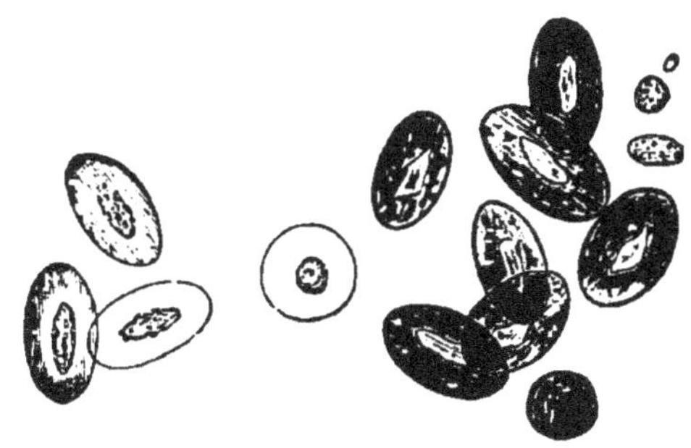

Fig. 28.
RED CORPUSCLES OF FROG'S BLOOD (400 Diameters).

3. We know by careful measurements the exact size of these globules in many animals. By examining a drop of blood with a microscope, we can often tell what animal it came from.

4. Such examinations are made of blood found on the clothing of persons accused of murder, and aid in determining their guilt or innocence. But this evidence should not be accepted as decisive. There are possibilities of mistake.

5. Besides the red corpuscles, there are also **white cor-**

puscles, a little larger, and spherical in shape, in the proportion of one white one to three or four hundred red.

6. The red globules have the property of attracting oxygen to themselves. They take it from the air in the lungs, and carry it to all parts of the body.

7. These globules give the blood its *red color*. If they were all taken out, it would be transparent and colorless. When there is a plenty of red blood in the blood-vessels, there is a rosy hue in the cheeks, and the lips are cherry red. This we call a healthy color. It is healthy because it indicates that there is blood enough, and that it is well supplied with globules. As these are the carriers of oxygen, a good number of them means plenty of oxygen in the tissues; and that is necessary to good health. On the other hand, colorless lips and skin indicate a lack of red globules, scanty oxygen, and ill health.

8. The watery part of the blood, in which the globules float, is called the **plasma.** This contains many substances dissolved in it, some of which are derived from the food, and nourish the body; and some are waste matters, which the blood is carrying away to be discharged.

9. When blood flows out of the blood-vessels, it soon thickens into a jelly. This is called the *coagulation* of the blood.

10. Blood does not coagulate in the blood-vessels during life if they are sound. If the blood is flowing very fast, it will not coagulate until the flow is checked. If the air is extremely cold, it does not coagulate quickly.

11. It is this coagulation of the blood that saves us from bleeding to death when we are wounded. The clots which form, stop up the mouths of the cut vessels.

12. *In case of a wound*, the blood should be helped to

coagulate by pressing on the spot so as to check the flow. Sometimes it is necessary to tie a handkerchief or a string around the limb above, or sometimes *below*, the wound. In this case the bandage should not be kept on too long, as other parts of the limb may suffer from want of blood.

13. *Loss of blood* causes great weakness, and at length fainting. When we faint, the heart almost stops its action, and the flow of blood becomes very slow. This gives it an opportunity to coagulate, and stop up the bleeding vessels. When the flow of blood can not be stopped, a fainting-fit may save life.

14. The body of a man contains *six or eight quarts of blood.* The loss of more than half of his blood would be certainly fatal, and the loss of a very much smaller portion might be so.

15. People near to death from loss of blood have sometimes been restored by throwing into their veins blood drawn from another person. This is called **transfusion.** It is a delicate and dangerous operation, and not often useful.

16. Some causes of impure blood are, —

1. Bad air.

2. Lack of exercise. This makes the flow sluggish. The organs whose work it is to purify it become inactive. Waste matter accumulates.

3. Too much or too rich food. The blood becomes loaded with matters which the system can not use and can not easily get rid of.

4. Too little or too poor food. The blood becomes thin, and unequal to the nourishment of the body.

5. Alcohol, which is itself an impurity, and unfits the blood for its work.

QUESTIONS.

SECTION I.—1. How does the body wear out?

4. How is it kept in repair? Why must it finally die?

6. Why do we need sleep?

7. Why is sleepiness a safeguard?

8. How can we know that we are not wearing out, when working very hard?

9. How much sleep does a grown person need?

10. Is sleep the only rest?

SECTION II.—1. From what materials is the waste of the body repaired? What is the most important?

2. How do we take in oxygen? How do we take in food and water? How are these distributed through the body?

3. What does the blood carry, besides the materials for repair?

SECTION III.—1. What is blood? What are blood-globules? What is their size? What proportion of the blood do they make?

2. What animals have red globules in their blood? What is their shape in man's blood? What is their shape in bird's blood. What animals have larger red globules in their blood than man?

3, 4. How can we distinguish the blood of different animals?

5. What globules are there, besides the red ones? How numerous are they?

6. What do the red globules do?

7. What makes the blood red? Why do we say that rosy cheeks are a sign of health?

8. What is the plasma of the blood? What does it contain?

9. What is the coagulation of the blood?

10. Does blood coagulate in the blood-vessels during life? What hinders coagulation?

11. What is the use of coagulation?

12. How should we treat a bleeding wound?

13. How may a fainting-fit be of advantage to a wounded person?

14. How much blood is there in the body? and how much may be lost without a fatal result?

15. What is transfusion?

16. Mention causes of impure blood.

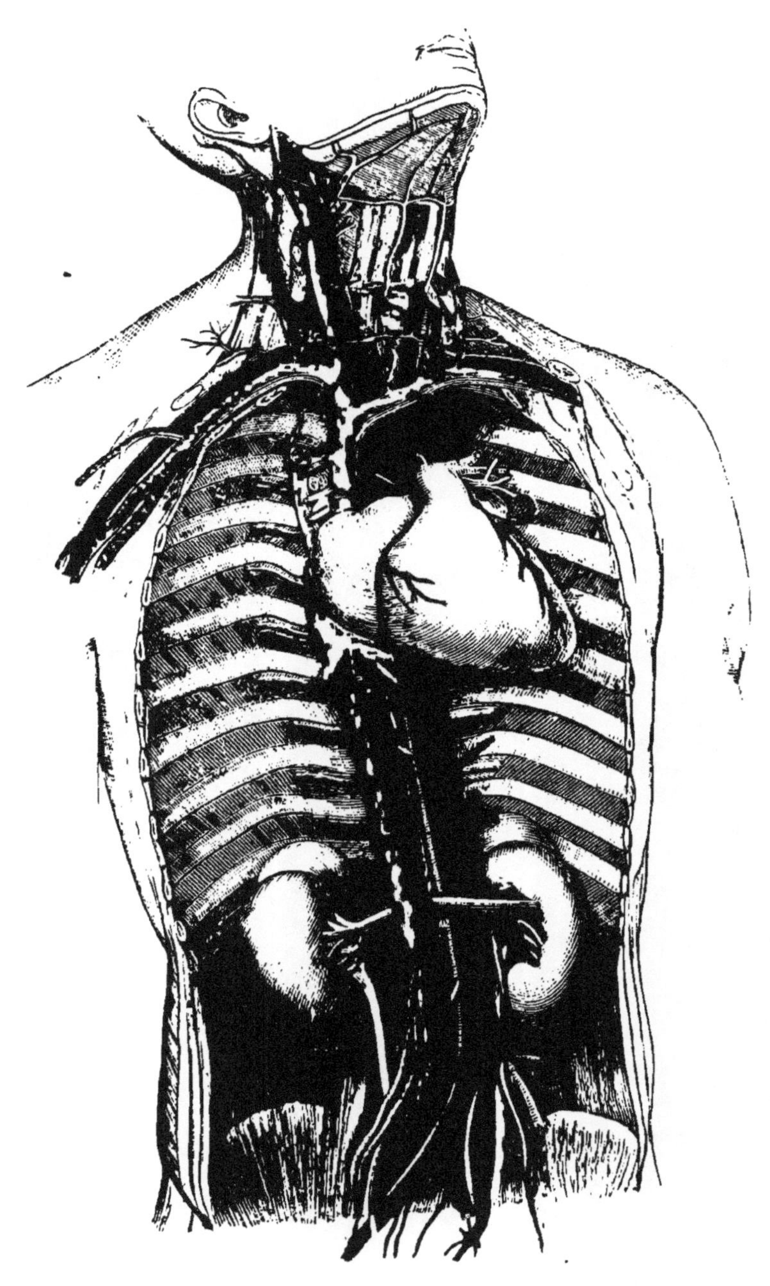

The figure on the opposite page shows the heart and the great blood-vessels.

The arteries are colored red, the veins are colored blue.

The right auricle and right ventricle of the heart are seen, and the beginning of the pulmonary artery, coming from the right ventricle. Arching from the heart, and passing down the back, is the aorta (red). It divides, at its termination, into the two common iliac arteries. The carotid arteries pass up on the sides of the neck. The subclavian arteries pass off to the right and left, and are continued as the axillary arteries. The venæ cavæ, ascending and descending, lie by the side of the aorta. The internal jugular vein accompanies the carotid arteries. The external jugular is just outside of the internal. The other veins have the same names as the arteries which they accompany. The intercostal arteries and veins run along the edges of the ribs on each side. The kidneys are seen, one on each side of the aorta.

CHAPTER V.

THE CIRCULATION.

SECTION I.—**1.** The blood is found in every part of the body. But the body does not hold it as a sponge holds water. It is rather like a house which has a supply of water carried through it in pipes. The blood in the body is all contained in pipes, called blood-vessels.

2. It is constantly in motion. Starting from *the heart*, it moves through the *blood-vessels* off to distant parts of the body, and then back to the heart again, making a circle. The heart and blood-vessels are, therefore, called *the organs of circulation.*

3. The blood-vessels are called **arteries, capillaries,** and **veins.** Coming out of the heart is a very large artery, called the **aorta.** This gives off branches as it passes on, and these branches, again, other branches, growing smaller as they divide, until at length the smallest branches are called capillaries.

4. The capillaries are very numerous, very small, and very close together. They form a net-work in every inch of bone and muscle and skin and brain,—a net-work finer than the finest silk. We can not put a fine needle-point into the skin without opening some of these capil-

---

SUGGESTIONS TO TEACHERS.—SECTION I. A beef's or sheep's heart, from the butcher's, will show all the parts named in the text. The action of the heart-muscle and of its valves may be partly illustrated with a Davidson's syringe.

laries, and drawing blood, so close together they are. If we could take out from the body of a man all the flesh and bone, leaving the blood-vessels, there would still be sufficient to form a perfect figure.

5. The *arteries* divide up into the *capillaries*, and the capillaries unite to form the *veins*. The little veins thus formed, unite to form larger veins, and so on, until at last they are all gathered into two large veins which enter the heart. These are called the **vena cava superior**, and the **vena cava inferior.**

6. The walls of the capillaries are very thin. Although there are no openings in them, a portion of the blood soaks through into the surrounding tissues; and, on the other hand, fluids containing waste matter soak into them, to be carried away.

THE HEART.

SECTION II.—1. **The heart** is made of muscle, and is hollow. It is the pump which keeps the blood moving.

2. It is situated in the chest, resting on the diaphragm, chiefly on the left side of the middle line. It is shaped like a pear, with the small end pointing down, and to the left.

3. It is inclosed in a sac called the *pericardium*. This sac is also pear-shaped; but the large end is on the diaphragm, and the small end points up.

4. The pericardium is lined with a very smooth membrane. The heart is covered with a continuation of the same membrane. This membrane is kept glossy and moist by a fluid that it gives out. When the heart moves in its sac, these two smooth membranes rub together without friction.

5. The *apex*, or point, of the heart is just beneath the fifth rib. Its *base* is a little to the right of the breast-bone. It extends up as high as the third rib. It is held in place by the diaphragm beneath it, and the large blood-vessels which run into and out of it. It is about as large as the fist, weighing from eight to twelve ounces.

6. We find, on examining the interior, that it is divided into *four cavities*. There is a division which runs across,

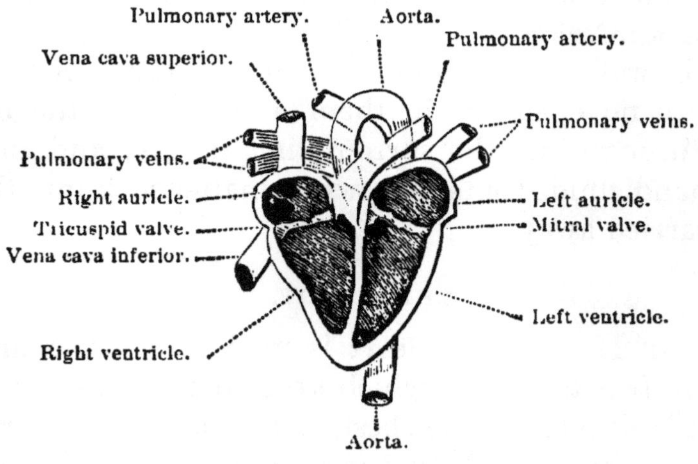

Fig. 29.
REPRESENTATION OF A SECTION OF THE HEART (Diagrammatic).

and one which runs lengthwise. This last divides it into right and left sides, the other into **auricles** (Latin, *auricula*, the external ear) and **ventricles** (Latin, *ventriculus*, the belly). These cavities are of nearly the same size, each holding about two ounces. The walls of the auricles are thin and loose. The wall of the right ventricle is about a sixth of an inch thick. That of the left ventricle is half an inch thick.

7. The two large veins which bring the blood from the

body back to the heart, open into the *right auricle.* Between the right auricle and the right ventricle, there is an opening about an inch in diameter. This opening is closed by the **tricuspid** (three-pointed) **valve.** This consists of three thin flaps, whose edges meet when the valve is shut. This valve opens from the auricle, and shuts so as to prevent the blood from passing back from the ventricle to the auricle. There are fine cords attached to the edges of these flaps, and to the wall of the cavity below, to prevent their shutting back too far.

8. Opening out of the right ventricle is the **pulmonary artery.** This soon divides into two, one of which goes to the right, the other to the left, lung. They there divide into capillaries, and the capillaries pass into the **pulmonary veins.** These are two in number for each lung. They empty into the left auricle.

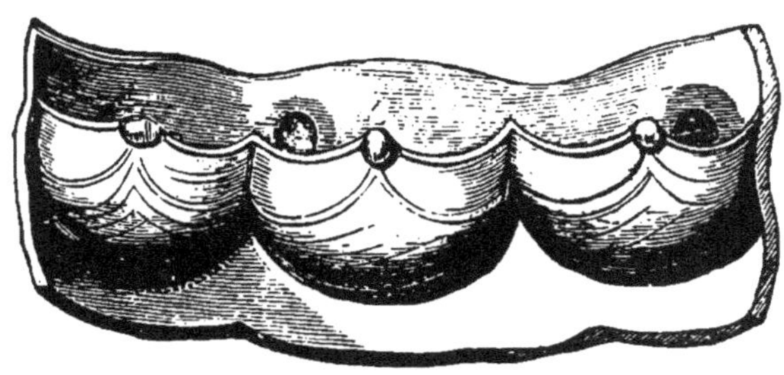

Fig. 30.
SEMILUNAR VALVES OPEN.

9. At the beginning of the pulmonary artery is a set of valves which prevent the blood from flowing back into the heart. These valves, called *semilunar valves,* are three pockets hung on the wall of the artery, their edges meeting all around. They open away from the heart; and, when the blood is going in that direction, they flatten up against the wall, just as a pair of doors will when a crowd is pushing through them. When the blood sets back toward the heart, it fills the pockets; and they bulge out, and together fill the tube,

and stop it up. These are very perfect valves. They are flexible, and float open like thin veils when the blood wishes to pass in the right direction; but they are strong; and, the harder it presses back, the tighter they fit.

10. The *left auricle* is similar to the right. The opening between it and the left ventricle is an inch in diameter, and is closed by the **mitral valve.**

Fig. 31.
SEMILUNAR VALVES PARTLY CLOSED.

The *mitral valve* differs from the tricuspid in having two flaps instead of three. When closed, it resembles a miter, or bishop's cap.

11. The *left ventricle* differs from the right only in having thicker walls.

12. Out of the left ventricle opens the great artery of the body, the aorta. Its entrance is guarded by semilunar valves precisely like those of the pulmonary artery described above. It gives off numerous branches, which go to all parts of the body, finally dividing into capillaries. From these capillaries the blood is returned by the veins to the right auricle.

13. The muscle of which the heart is composed is peculiar. It is striped, and yet not voluntary. As it must act during sleep, it is made independent of will or thought.

The heart and the blood-vessels are lined by a membrane as smooth as satin.

ACTION OF THE HEART.

SECTION III.—1. The heart is, in the body, what the mainspring is, in a watch. Like all other muscle, it has the power of contracting. By constant and regular contractions it keeps up the circulation, and thus sustains life.

2. The heart-muscle is remarkable for its endurance. No other muscle could do its work. Through a whole lifetime, sometimes a hundred years, it never pauses for one minute.

3. It might seem, therefore, to be an exception to the general law, that rest is necessary for all organs; but, after each contraction, it has a very short time of relaxation and rest. This time is not more than two-fifths of a second; but, as it comes every second, its whole amount in twenty-four hours would be eight or nine hours. Moreover, in sleep the heart beats less rapidly, and is not obliged to make the special efforts which are so often required of it during the day by rapid movements or excitement.

4. The heart is, doubtless, tired after labor, and contributes to the general sense of fatigue; but in health we have no special feeling of the heart. Great and long-continued care such as business men are often subjected to, or protracted muscular exertion like that of soldiers on a march, sometimes so exhausts the heart, that it acts irregularly, and feels distress. This condition, known to physicians as "irritable heart," may be brought on by prolonged dissipation.

5. The number of heart-beats in a minute varies at different periods of life. In an infant it is one hundred and twenty or more; in a child under fourteen, eighty or more; in a grown person, about seventy-two. But it may be ten beats more or less; and, in rare cases, there is a still wider variation in health.

6. Many things cause a temporary variation. It is less in sleep: it is greater in active exercise. One reason why we soon become exhausted by running, is that the heart is stimulated to such rapid action. It is greater after eat-

ing in moderation. *Excitement* of any kind increases the number and force of heart-beats so that they can be felt, and sometimes heard. In the affection called palpitation, the action of the heart is unnaturally rapid, and sometimes hard enough to shake the body.

7. The heart has been called a pump. It is really a double pump. There are two streams flowing out of it with each contraction, and two streams flowing into it with each relaxation. The two sides of the heart have no direct communication, and are often spoken of as the right and left heart, as if they were separate.

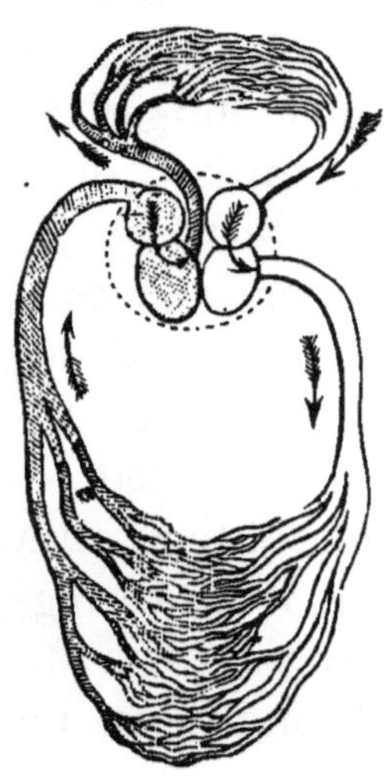

Fig. 32.
REPRESENTATION OF THE GREATER AND LESSER CIRCULATION.

8. There are, accordingly, two distinct circles of blood. One begins at the right ventricle, goes through the pulmonary artery to the lungs, and back through the pulmonary veins to the left auricle. This is called the **pulmonary circulation,** or the **lesser circulation.**

The other begins at the left ventricle, and goes through the aorta, to be distributed to all parts of the body, and, passing through the capillaries and veins, is poured into the right auricle by the vena cava superior and the vena cava inferior. This is called the **systemic** or **greater circulation.**

9. Let us observe a contraction and its effect. Suppose the auricles to be full. The blood has been pouring into the right one from the venæ cavæ, and into the left from the pulmonary veins. Now the auricles contract.

Their contents can not go back into the veins, for they are full; but the ventricles have just emptied themselves, and are opening for a new supply. The blood is therefore forced suddenly on through the passages, with their open valves, into the ventricles. Directly the ventricles, now full, begin to contract. The blood sets back against the valves it has passed, and shuts them tight, just as a crowd trying to get through a door which opens towards them will often close it: but the passages into the pulmonary artery and the aorta are clear,— their valves (*the semilunar*) open out; into them the blood pours, the ventricles still contracting until all is squeezed out. Then the blood in the arteries sets back, and shuts the semilunar valves; and the ventricles relax, and open for another supply.

SOUNDS OF THE HEART.

SECTION IV.—1. If the ear be placed upon the chest over the heart, two sounds can be distinctly heard, repeated with each beat. One is quickly followed by the other, and then there is an interval. The second is shorter, and higher pitched than the first. They may be indicated by the signs (lub - dub.) The first sound is caused chiefly by the closing of the tricuspid and mitral valves, as the ventricles contract. The second sound is caused by the closing of the semilunar valves after the blood has passed into the pulmonary artery and aorta. When the heart is diseased, these sounds are changed; and the changes in sound indicate to the ear of the physician the particular changes in the heart.

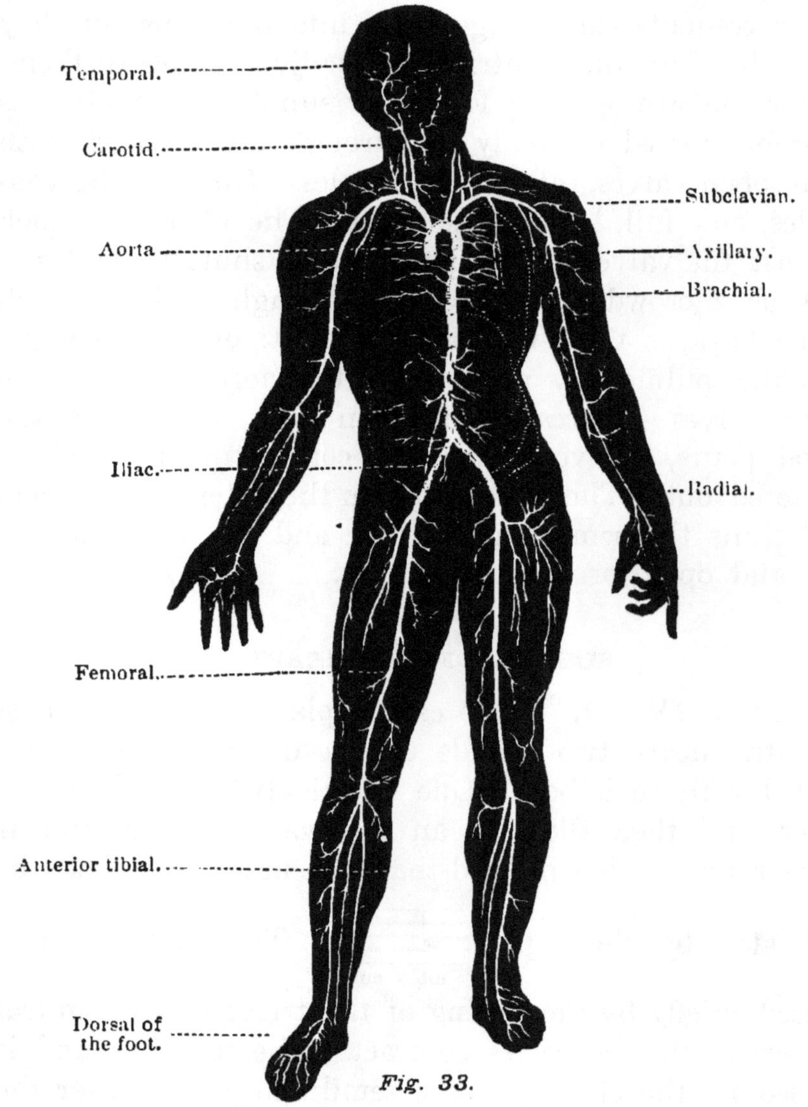

Fig. 33.

REPRESENTATION OF THE ARTERIES.

THE ARTERIES, VEINS, AND CAPILLARIES.

SECTION V.—1. The aorta arches from the base of the heart across to the backbone, by the side of which it descends. On a level with the top of the hip-bones, it divides into two vessels, one of which supplies each of the lower limbs. From the arch, branches are given off, which supply the head and the arms.

2. In the thigh, the large vessel which carries the main stream is called the **femoral artery.** In the leg it is called the **tibial artery.**

3. The **carotid** arteries carry the blood from the arch to the head. They pass up on each side of the neck, and their throbbing can often be seen. The **subclavian** arteries lie behind the collar-bones. They extend to the armpit, and in that situation receive the name **axillary.** From the armpit to the elbow they are called **brachial.** At the elbow they divide into the **radial** and **ulnar.** The radial lies on the thumb-side of the fore-arm, and is the one in which the pulse is commonly felt.

4. All the large arteries lie deep. They can be felt or seen on the surface, only in a few places. But many large veins lie just beneath the skin, and can be traced for some distance.

The deep veins run by the side of the arteries.

5. The veins frequently have the same names as the arteries which they accompany. Sometimes they have special names. The large veins of the neck are called **jugular** veins. The **superficial jugular** is the prominent vein just beneath the skin on each side.

6. The blood flows in the large arteries much more rapidly than in the small ones. Its flow is slower in the veins than in the arteries. It is slowest of all in the capillaries. This is for the same reason for which a stream flows more slowly when its channel is wide than when it is narrow. The capillaries are very small; but there are so many of them, that their total blood-channel is really wider. It is estimated that it is three hundred times as wide as the aorta.

7. Though many men of great minds gave much study to the human body, it was not learned until less than three hundred years ago that the blood circulated. William Harvey, an English physician, made the discovery, and gave it to the world.

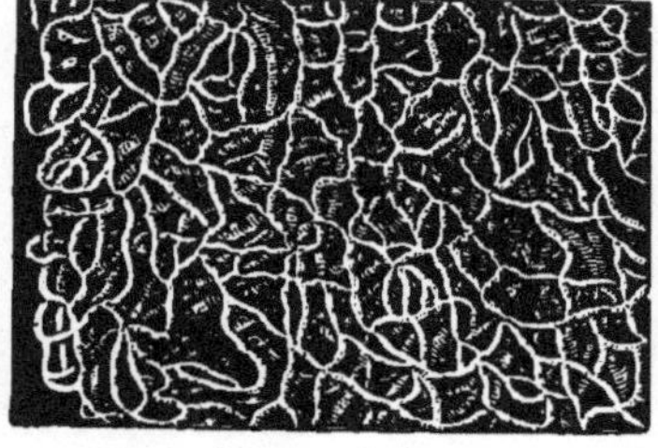

Fig. 34.
CAPILLARY PLEXUS MAGNIFIED.

8. The *walls of the capillaries* are a single layer of a thin membrane. Through them nutritious fluids soak out, and waste fluids soak in.

9. Some points in which arteries and veins differ:—

1. In the direction of their current. In arteries it always runs from the heart: in veins it always runs toward the heart.

2. In position. We do not find large arteries running just beneath the skin, as veins do.

3. In color. The blue tint of their blood shows through the walls of the veins.

4. In thickness. The arterial walls are thicker and firmer.

5. The veins have *valves* at intervals in their course, which allow the blood to move forward, but not backward.

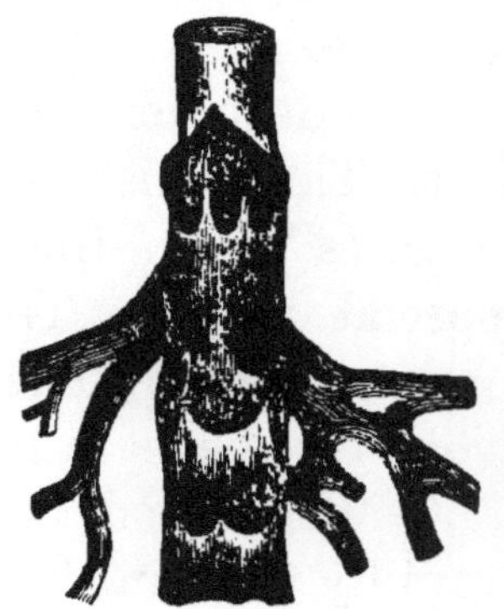

Fig. 35.
VEIN LAID OPEN, SHOWING THE VALVES.

10. **In case of a wound,** we can tell whether the blood is from an artery or a vein—

1. By its color. Venous blood is dark: arterial blood is light.

2. By the manner of its flow. Venous blood flows in a steady stream, arterial blood in jets.

11. The flow of blood from a vein stops more readily than that from an artery. A bandage is commonly sufficient. The arterial flow is more forcible. When the blood spurts, it must be checked by firm pressure on the vessel until a physician arrives. It may be necessary to tie the artery.

The pulmonary artery is peculiar because it carries blue venous blood. The pulmonary veins carry red arterial blood. In these respects they are exceptions to the general rule.

12. Arteries and veins are very strong. In experiments made, they have borne a pressure equal to a weight of a hundred pounds without bursting. But they sometimes burst. In old age they become brittle. Certain diseases make them brittle.

THE PULSE.

SECTION VI.—**1.** At a certain spot in the wrist, we can feel a regular beat, which we call the pulse. A pulse may be felt in several other places, but the wrist is the most convenient. The pulse is a sudden rising-up of the wall of the artery under the finger, and a slight stirring of the whole artery in its bed.

2. Every time the heart contracts, four or five ounces of blood are thrown into the arteries. These, being elastic, stretch to receive it. Between the beats they contract, forcing the blood along into the capillaries.

This stretching is greatest in the aorta, close by the

heart, and diminishes as the vessels grow smaller. By the time the capillaries are reached, there is no stretching, hence no pulse. The veins have no pulse.

3. The pulse tells us,—

1. How fast the heart is beating.

2. How strong its beat is.

3. How firm the coats of the arteries are. The last fact is important. The arterial walls are sometimes flabby, and sometimes hard and resisting. Their condition indicates the state of the system.

VARIATIONS IN BLOOD-SUPPLY.

SECTION VII.—1. You might suppose, from what has been said of the circulatory apparatus, that the same blood-vessels are always of the same size, and that the blood-supply to the same part is at all times of the same amount. This is not the case. Although there are no stopcocks in the blood-vessels, there is a contrivance of Nature by which the size of the blood-vessels and the amount of blood in any part is increased or diminished according to the need of the moment.

2. In this way *the blood is partly shut off* from the brain when we need sleep; it is turned on to the stomach during digestion, and to the muscles in active exercise; it is shut off from the skin when it is cold, and sent there to be cooled when we are warm.

3. The walls of the arteries consist partly of fibers of involuntary muscle running crosswise. When these muscle-fibers contract, they make the artery smaller: when they are relaxed, it is larger. They are not under the control of the will, but are influenced by special causes. For example, the arteries in the coats of the stomach will

grow larger when food touches it. The arteries of the salivary glands will grow larger, and fill with blood for the manufacture of saliva, when we smell food, thus "making the mouth water." The arteries of the skin will grow larger, and fill with red blood, when exposed to heat. The arteries of the face will expand, and cause a blush, when we feel shame.

4. The apparatus which regulates the size of the arteries belongs to the nervous system.

EFFECT OF ALCOHOL ON THE CIRCULATION.

SECTION VIII.—1. *Alcohol* in moderate amount makes the heart beat faster. In health it beats fast enough; and the extra beats, which amount to a good many in a day, are labor lost. Worse than this, they are wearing out that patient organ.

2. *Alcohol* has the power to change muscle gradually into fat. The heart is particularly liable to this change. As the fibers soften down, it loses its strength. It can not do the work of pumping the blood through the body with its natural vigor. The body, therefore, suffers in every part. Such a fatty, soft heart is liable to break suddenly.

3. As old age comes on, the walls of the arteries frequently become changed in previously healthy persons. They get fatty and soft, or chalky and brittle, in spots. In this state they easily burst, if by excitement, or overeating, or some other cause, they are unusually full. This is especially liable to occur in the blood-vessels of the head. The blood pours out; and, as the skull is a tight box, the brain is pressed so hard that it can not act. The person so affected becomes unconscious, and is very

likely to die. This takes place suddenly, and it is called **apoplexy.**

4. *Apoplexy* is a disease of advanced life. These changes do not take place in the arteries of the young without special causes. *Alcohol* is such a cause. It makes the young man old before his time, and liable to sudden death by the rupture of an artery in the brain.

QUESTIONS.

SECTION I.—1. In what way is the blood held in the body?

2. Why are the heart and blood-vessels called organs of circulation?

3. What are the three kinds of blood-vessels?

4. Describe the capillaries.

SECTION II.—1. What is the heart?

2. Where is it situated?

3. What is the pericardium?

4. How is friction from the movements of the heart avoided?

5. Where is the apex of the heart? How high does it extend? How is it held in place? How large is it?

6. How is its interior divided? Name its cavities. What is the size of each? What is the thickness of their walls?

7. Describe the passage between the right auricle and ventricle, and its valve. What vessels open into the right auricle?

8. What vessel opens out of the right ventricle? What is its course? What vessels empty into the left auricle?

9. Describe the semilunar valves of the pulmonary artery.

10. Describe the passage between the left auricle and ventricle, and its valve.

11. What is the difference between the left and the right ventricle?

12. Describe the aorta and its valves.

13. Describe the muscle of the heart.

SECTION III.—1. What does the heart do?

2. Does the heart get tired?

3. How does the heart rest?

4. How many times does the heart beat in a minute?

5. What causes make the number of beats in a minute greater or less?

6. Why may the heart be called a double pump?

7. What is the lesser circulation? What is the greater circulation?

8. Describe in full a contraction of the heart.

SECTION IV.—1. What sounds does the heart make, and how?

SECTION V.—1. Describe the course and termination of the aorta.

2. Where is the femoral artery? the tibial?

3. Where are the carotids, the subclavian arteries, the axillary, the brachial, the radial, the ulnar?

4. Do large arteries run as near the surface as large veins?

5. How are the veins named? Where are the jugular veins?

6. Where does the blood flow fastest? Where slowest?

7. Who discovered and proved the circulation of the blood?

8. What are the walls of the capillaries?

9. Name some points in which arteries and veins differ.

10. How can we tell, in case of a wound, whether the blood comes from an artery, or from a vein?

11. Is the flow from a vein as easily stopped as that from an artery?

12. Are veins and arteries strong?

SECTION VI.—1. What is the pulse?

2. Is there any pulse in the capillaries? in the veins? How does the blood flow from an artery? How from a vein? How from the capillaries?

3. What does the pulse tell us?

SECTION VII.—1. Is the supply of blood to the same part always the same?

2, 3, 4. How is it made greater or less?

SECTION VIII.—1. What is the immediate effect of alcohol on the heart?

2. What change does alcohol sometimes cause in the substance of the heart?

3. What change does alcohol sometimes cause in the blood-vessels? What is apoplexy?

4. How does alcohol cause it?

CHAPTER VI.

FOOD AND WATER, STIMULANTS AND NARCOTICS.

SECTION I. — **1.** By **food** we mean all substances that we eat or drink to satisfy hunger and nourish the body.

2. If we have to repair any manufactured article, we use the same kinds of material that the article was made of. We repair a harness with leather, a stove with iron, a table with wood. So our food, which repairs the body, must contain the same substances that the body contains.

3. Analysis by the chemist shows that the body of man consists of *fifteen substances, called* **elements.**[1]

Elements.	Parts in 100.	Elements.	Parts in 100.
Oxygen	72.	Sulphur	.1476
Hydrogen	9.1	Sodium	.1
Nitrogen	2.5	Potassium	.026
Chlorine	.085	Iron	.01
Fluorine	.08	Magnesium	.0012
Carbon	13.5	Silicon	.0002
Phosphorus	1.15	Manganese	A trace.
Calcium	1.3		

4. Our food must contain every one of the above elements. Of some, the body contains little; and we require

[1] These substances are not always in exactly the proportions given in the table. Each would differ in amount in different bodies, and in the same body at different times.

but little in our food. *Sulphur* is one of these. Others are abundant in the body, and are found in almost every article of food. *Carbon* is one of these. If any of these elements is entirely wanting in our food, we suffer, and would starve to death for lack of some of them. If we should try to live on food which contained no *phosphorus*, for example, we should become diseased, and die. Almost every thing that we eat contains phosphorus.

5. Other elements than the fifteen given above, are not required in our food. Silver, for example, does not form part of the body, or of our food; but nitrate of silver is sometimes used as a medicine.

6. All of these fifteen elements are found in the *air*, the *earth*, and the *water*. But animals can not feed on earth and air and water. **Plants** can, and that is one of the great distinctions between plants and animals. Four-fifths of the air is nitrogen. But men, and other animals, would die for want of nitrogen, even while they were drawing it into their lungs with every breath, if they could not get food containing it.

7. It is *the work of plants* to take the elements, and make them into food for animals. The plant feeds on earth, air, and water. This earth, air, and water become a part of itself. The animal feeds on the plant, or on other animals.

8. Probably every plant is food for some animal. But there are very many which are not food for man. Some are poisonous. A large number contain the necessary elements, but can not be digested by the human organs of digestion. A fertile prairie which would fatten a herd of buffaloes would starve a man; because the grasses, though not poisonous or distasteful, contain too much indigestible matter for his stomach.

9. Some animals live only on other animals. Their teeth are not fit for chewing vegetable food Other animals live wholly on plants. Their teeth are made for grinding, but not for tearing flesh. The cow, for example, has no cutting-teeth in her upper jaw. She has broad grinders. Man eats both animal and vegetable food. His teeth are adapted to both.

10. The infant, and the young of many of the lower animals, live on *milk.* No other article of food so well combines all the necessary elements as this. It is easily digested, and, after infancy is passed, is still an excellent article of diet. In sickness it is often used with advantage as the sole nutriment.

11. Man's ordinary diet consists of meats (including fish and eggs), starchy foods, sugar, and fat.

MEATS.

12. Of **meats,** there are many kinds. The appetite is gratified, and the body better nourished, by a variety. *Beef* is the best. *Pork* is the staple meat-food of large numbers of people It is not so wholesome as beef, for two reasons: —

1. It commonly contains a great deal of *fat,*—too much for constant use.

2. It is more likely than beef or mutton to contain the young of the tapeworm, and other *parasites.*

Veal is tender and good, but not so easily digested as beef, nor is it so nourishing. *Lamb* is very easily digested. *Mutton* is more nourishing than either lamb or veal, and is nearly as good food as beef. To some stomachs, however, it is not acceptable. Each of the various kinds of *game* has its peculiar flavor, but they do not differ much in nutritive value.

All meat is better and more tender for being kept for a time after killing.

13. *Fish*, on the other hand, is better when perfectly fresh. It does not differ very much from flesh in its chemical composition; though it contains more water, and less fat. It is lighter and less stimulating than flesh.

14. Of *shell-fish*, oysters occupy the first place. They are palatable, light, and nourishing. They tempt the appetite of an invalid without distressing his stomach. In spring and early summer, they should not be eaten.

15. *Lobsters*, *crabs*, and *shrimps* are less digestible. Vinegar and stimulating spices are commonly added to them, which spur up the stomach to its task. They should be avoided by invalids.

16. *Eggs* contain much nourishment, solidly packed away for the support of the young fowl before he breaks out of his shell. They are palatable, and easily digested.

STARCHY FOODS.

17. The **starchy foods** include all the grains — wheat, oats, corn, etc. — and vegetables.

Dried *wheat* contains, in 100 parts, 66 parts of starch.

Dried *oats* contain, in 100 parts, 60 parts of starch.

Dried *rice* contains, in 100 parts, 88 parts of starch.

When we remember that many millions of the human family live chiefly on rice, whose solid substance is almost all starch, we have good proof of the importance of starchy food.

18. Besides starch, the grains contain *mineral matters*, *fat*, *sugar*, and a substance, similar in chemical composition to meat, called *gluten*.

Wheat is, on the whole, the most valuable of the grains

as food; and men prefer it when they can get it. But each grain has its own advantages. Corn is rich in oil. Oats have much mineral matter, and much fibrous, branny substance, which make them indigestible for some delicate stomachs. For the majority they are healthful. It has been claimed that ground wheat is better food with the bran in it,—in which condition it is called graham flour,—than when the bran has been separated from it, and it has been made into fine flour. In sifting out the bran, we take away a good deal of nutritive substance; and what is left is chiefly the starch. The advocates of this view were numerous a few years ago, and were called *Grahamites*, after their leader Graham. Nevertheless, fine white flour is still the choice of the majority. It is true that much nourishment is lost in the bran. But it is so mingled with woody matters, that it is not easily extracted by the stomach. For healthy people, it is better to use white flour, and to get, in other and more digestible foods, what is contained in the bran. But a stomach that is sluggish in its action, is stimulated and aided by the presence of the bran. For such, graham flour is excellent food.

19. The *potato* is the most popular of all vegetables, though it has not been in common use more than three hundred years. It has been estimated that it forms at least three-fifths of the food consumed in Ireland. Its chief solid ingredient is starch. Much of the starch used in the laundry is extracted from potatoes. It is superior to other vegetables in being drier, and containing less fibrous substance.

20. *Pease* and *beans* contain very little water, and a great deal of solid matter. They are very nutritious, and are most valuable for feeding armies, and other large

bodies of men, because they contain so much in small bulk. But, because they are so solid, it is hard for the stomach to digest them.

21. *Turnips*, *beets*, *cabbages*, and other garden vegetables, contain more water and less nutritive matter than those already referred to. They form an agreeable and healthful addition to more nutritious articles.

SUGAR.

22. There are many kinds of *sugar*, all sweet, but differing in taste and solubility. Most of our vegetable food contains sugar, and a great deal of pure sugar is used in addition to this. It is especially pleasing to the taste. The fact that children are so fond of it, is a proof that it has important uses in the growing period. But, if too much is eaten, it spoils the appetite and digestion, and injures the health.

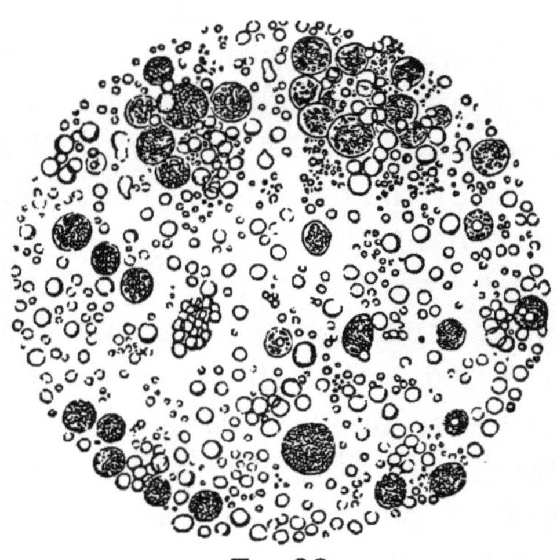

Fig. 36.
A DROP OF MILK MAGNIFIED, SHOWING THE OIL-GLOBULES.

FAT.

23. Most vegetable, as well as animal, foods contain some *fat*. It is customary to add oils, or fats of some kind, to other food. It is itself nutritious, and aids in the digestion of other things. *Butter* is the oil of milk. If we look at a drop of milk with a microscope, we see innumerable little particles of oil. Oil is lighter than water;

and, therefore, if milk is allowed to stand, the oil will rise to the surface, making the cream. After the cream is taken off, the milk looks thinner and more transparent. If all the cream could be removed from it, it would be clear, like water. In making butter, the milk is churned until all the oily particles are stuck together in a mass, and the watery part is separated: this latter is called buttermilk.

COOKING.

24. Animals eat their food raw. Men, even the most uncivilized, cook much of it. *Cooking* is useful in three ways.

1. It renders food more *digestible.*

2. *It kills parasites* in meats, and stops putrefaction.

3. *It creates flavors* which are attractive, and stimulating to the digestion.

25. *Raw meat* is not unpalatable: a fondness for it is often acquired. A disease called *trichinosis* is caused by eating pork, which contains a minute worm called a trichina. This worm can not be recognized in pork without a microscope. It is killed by cooking: smoking is not sufficient. The disease is, therefore, found only among those who have eaten raw pork. Tape-worms are also acquired by eating raw pork and beef.

26. We can not digest starchy foods, such as the grains and vegetables, in sufficient quantity, unless we cook them. By cooking, we soften them so that they can be easily separated into fragments, and dissolved by the digestive juices. Besides, we add to them, in cooking, various articles, such as salt, fat, sugar, and spices, which improve their taste. Flour is, in this way, made into **bread**, and cake, and pastry of many kinds.

27. The essential points for making good bread are,—

1. Good flour.

2. Thorough kneading, to mix all the ingredients.

3. Good yeast.

4. Good judgment exercised in keeping the rising mass just warm enough, and letting it rise just long enough.

5. Having the oven *just hot enough*, and taking out the bread at just the right time.

28. Good bread is light and sweet. Bread is *light* when the carbonic-acid gas formed in the fermentation caused by the yeast has penetrated the whole loaf, making innumerable holes and pores in it. When such bread is eaten, the digestive juices easily enter these pores, and spread through and act upon every part.

Bread is *heavy* when carbonic-acid gas has not penetrated the mass, either because the yeast is poor, or the flour is poor, or because it has not been thoroughly kneaded. It makes a solid lump in the stomach, which the digestive juices can not easily enter; and that organ becomes weary and sore in struggling with it.

Bread may be *sour*, either because there is too much yeast in it, or because fermentation has gone on too long. In that case, acetic acid is formed in it.

29. *Pastry* and *cake* are not so wholesome as bread. They please the taste, and in small amounts are not injurious. Pie-crust is not commonly so light as bread. Besides, it has a good deal of lard or butter thoroughly mixed with it. This fatty matter, which is not acted upon by the juices of the stomach, coats over the particles of the flour, and prevents the gastric juices from reaching them; so they must pass out into the small intestine undigested. This is the reason why all *food fried in fat* is less easily

digested than food cooked in other ways. In the process of frying, the fat is so thoroughly mixed with the food, that it renders it partly proof against the stomach-juices.

Cake contains too much sugar and butter, in proportion to its other nutritive matters, to be wholesome in large quantities.

30. *Cookery* is an art which is very important to health and comfort. It must be learned by practice. But those who understand the principles of chemistry and physiology, on which it rests, will acquire it more readily, and be more completely masters of it.

MINERAL SUBSTANCES.

31. *Mineral substances* are mingled with all our food. *Salt* is the only solid mineral matter that we take pure. It enters into every part of the body, solid or fluid, and aids the processes of life.

Mineral matters are not commonly changed by digestion.

WATER.

32. *Water* forms about seventy of every hundred parts in the body. It must be constantly supplied, therefore, to make up for the waste of the parts. Digestion, absorption, and circulation would stop without water. The craving for it is stronger than for food.

33. Waters used for drinking always contain a small portion of mineral salts, of gases, and of vegetable matter. Water which is absolutely pure—as only distilled water is—is flat and tasteless. The mineral matters in drinking-waters are such as are not harmful to the system, unless there is too much of them. In that case, they are irritating to the bowels and kidneys. When water is

carried through *lead pipes*, it sometimes dissolves enough of the lead to become *poisonous*. Whether it will do this, or not, depends on what it already contains. Some waters may be carried through lead pipes with perfect safety, either because the water does not act on the lead at all, or because it contains certain mineral matters which form a crust, lining the pipes, and protecting them from further action. When lead pipes are used, the question whether the water acts upon them, should be settled by a chemist if necessary. Galvanized iron or tin pipes are safer. Water that is constantly running through the pipes is less likely to contain much lead than that which stands still in the pipes for long periods. Lead, when taken in small doses in this way, produces its effects very gradually; and the health is often seriously affected before the cause is discovered. Among the symptoms of lead-poisoning are colic, and paralysis of certain muscles.

34. *Offensive* and *poisonous matters*, animal and vegetable, sometimes find their way into drinking-water, and, being dissolved in it, give no sign of their presence. *Sewage*, and the germs of disease, may thus be taken in. It is necessary, therefore, to guard the well, or water-pipes, very carefully from all impurities. Wells and reservoirs are often placed where they catch the drainage from barnyards, or other receptacles for filth. Such drainage will go through the soil much farther than is commonly supposed. The fact that water is clear and sparkling and odorless does not prove it pure. A well or reservoir should not be located within thirty feet of any filthy spot. Even at that distance it is not safe if the ground is porous, and slopes toward the well.

Aqueduct-pipes sometimes become leaky, and draw in filth. Constant watchfulness against the foes to health, which would enter in this way, is necessary.

EATING AND DRINKING HABITS.

35. The eating and drinking habits of mankind vary greatly. They are modified by climate and by surroundings. The Esquimaux drink fish-oil and eat candles with a relish; the Hindoo lives upon rice; the Arab supports life, and performs great journeys, on a handful of grain a day. The European or American requires more and better food. The human body can adapt itself wonderfully to its circumstances. But those nations, which, by reason of their geographical situation and their wealth, have been able to obtain the best and most varied diet, have the best and strongest bodies.

36. There is also a great variety in the habits and tastes of members of the same race or community. Some prefer one kind of food, and some another. Some eat two meals, and some three. Constitutions, habits, and circumstances make great differences. One man may thrive on food that would destroy another. A brain-worker may accomplish most and feel best if he eats little until noon: a day-laborer would lose his vigor under such a practice. One man is over-stimulated by a meat-diet: another ought to live chiefly on meat.

If Nature had not made mankind capable of such variations in habit, the work of the world could not be done.

37. A *healthy appetite* is Nature's guide to right habits of eating and drinking, but Nature intended that appetite should be controlled and regulated by reason. Each man will thus adopt that course which is best for *him*.

38. When large bodies of men have to be fed, as in the army or navy, it becomes necessary to find out just how much of each kind of food a man requires daily. By combining physiological reasonings with experiment, Professor Dalton found, that for a man in health, taking free exercise in the open air, the following was a sufficient daily ration: —

Meat	16 ounces.
Bread	19 "
Butter	$3\frac{1}{2}$ "
Water	52 "

Total, water, $3\frac{1}{4}$ lbs.; solids, 2 lbs. $6\frac{1}{2}$ ounces.

Men at hard labor require more, and those who are entirely inactive, less.

STIMULANTS AND NARCOTICS.

SECTION II. — 1. *Stimulants* are substances which excite. *Narcotics* are substances which benumb and stupefy. Some substances are both stimulants and narcotics. *Alcohol*, if taken in small quantity, is a stimulant: in large quantity it is a narcotic.

2. Nature supplies us with certain food-stimulants which are useful. Stimulant is from a Latin word, *stimulus*, meaning a goad. These substances afford little or no nourishment, but they goad the appetite and the digestive organs to greater activity. Such are pepper, spice, and mustard. Though good in moderation, they may be used in such quantities as to injure the stomach, and make the body liable to various disorders.

TEA AND COFFEE.

3. *Tea* and *coffee* are used by all civilized nations. Their immediate effect is cheering. They often aid digestion. They satisfy the cravings of the stomach, and enable men to endure hardship better. But many persons are injured by them,—some, because they drink too much; others, because they are too susceptible to their action. The morning cup of coffee is often paid for by a daily headache. There are intemperate tea-drinkers. Tea contains a good deal of tannin, which has a tendency to check the action of the stomach and bowels. Tea made as the Chinese make it, by simply pouring boiling water on the leaves, and not steeping long, contains more of the delicate flavors, and less of the tannin, than that which is boiled for some time. The fact that some healthy people can not go to sleep at the usual time if they take a cup of tea in the evening, shows that it has a decided effect on the nerves. Fretfulness and irritability, and palpitation of the heart, are results of its immoderate use.

4. *Coffee*, as ordinarily made, contains more solid matter than tea, and is, therefore, more nourishing. It has less of the astringent principle, and is more soothing. Both tea and coffee will sometimes relieve headache.

It is necessary that the stomach should be very warm in order to digest well. If food is taken cold, or if the whole body is in a chilly state, the cup of tea or coffee awakens the stomach to activity by conveying heat in an agreeable form.

5. *Growing children* should drink neither tea nor coffee. Their fresh and vigorous bodies need no other stimulants than air, exercise, and simple food. They are,

besides, much more susceptible than grown persons to the bad effects of these things.

TOBACCO.

6. *Tobacco* is a drug of very great power. A drop of the oil extracted from it, placed on the tongue of a dog, will kill him almost as quickly as prussic acid. It is occasionally used as a medicine, but with caution, because its effects are so severe.

Its odor and taste are disagreeable. When taken for the first time, it causes intense nausea and wretchedness. By persisting in its use, the revolt of nature is commonly overcome, and a liking for it is acquired. Its immediate effect is then a feeling of tranquillity and comfort. Soldiers and sailors, and others who have to endure great physical labors, find in it support and relief. Students use it because it gives them a freer flow of ideas; men of pleasure, because it causes agreeable sensations. In some cases it seems to do no harm: in others, its bad effects are easily seen.

7. It *diminishes the appetite* for food. It sometimes *causes disease* of the mouth and throat. It *weakens the stomach*. It becomes an absolute necessity: brain and stomach demand it. Often the amount used must be increased until it is hardly ever out of the mouth. At first soothing the nerves, it at last makes them irritable and unsteady.

8. Its effect on the heart is so marked, that the term "*smokers' heart*" is well known to physicians as indicating irregular and weak action. The student whose brain is rendered more active by it, may find himself suffering at length from head-troubles, and failure of nerve-strength;

and he ends his work before his time. The business-man who has gone beyond his natural powers, with the aid of tobacco, finally breaks down entirely.

9. It is evident that the effects of tobacco on *the young* are especially evil. Boys who use it are dwarfing their minds and bodies. They are so changing the system from its natural healthy condition, that it is preparing for disease, and acquiring tendencies that lead to dissipation and worthlessness. They have no possible excuse for its use.

10. Whatever satisfaction it may give, is purchased at the expense of slavery to it, often of personal neatness, and at the constant risk of offending companions unused to it. In some occupations it is a serious hinderance to success.

OPIUM.

11. "*Opium*," said the great physician Boerhaave, "*is the finger of God.*" Whoever has seen or felt the cessation of pain that seemed unendurable, under its power, can respond to his sentiment. Rightly used, it is a boon: perverted to purposes of sensual gratification, it is the cause of untold misery.

12. Opium not only relieves pain, but, in small doses, gently stimulates the brain and nerves, making the taker able to endure and accomplish more than he otherwise could. In larger doses, it induces a dreamy state, in which he is released from the annoyances of life, and wanders freely on the wings of imagination. In poisonous doses, a stupor comes on, in which the breathing becomes slow, sometimes not more than two or three breaths being taken in a minute. The pulse is also slow and full. When this point is reached, the slumber is likely

to become deeper until death ensues. To avoid this, it is necessary to do every thing to keep the drowsy person awake, —to beat and pinch him, to keep him on his feet, and walking, to throw cold water on him.

13. Those who acquire the **opium habit** become enslaved to it. Its chains are even stronger than those of alcohol. The misery which the attempt to go without occasions, overcomes the strongest will. The victim believes that he will die, that he is dying, and that only opium will save him; and, in these circumstances, those apparently most conscientious will lie and deceive to obtain it. The practice tends to kill truthfulness in the soul, and to undermine the whole character.

14. Opium has a paralyzing *effect on the digestive apparatus.* It checks the flow of digestive juices, and the action of the muscular walls of the bowels. It takes away the appetite for food. In those not habituated to it, it commonly causes nausea. When not under its influence, the opium-taker suffers from headache and depression. His nerves are relaxed, his mind dull, and his will feeble. He is unfitted for the work of life, and his only object is to gratify his craving.

15. Opium should never be taken, except *under the direction of a physician.* Those who are suffering from pains which are likely to return and visit them frequently, should avoid it. It is better to endure pain than to become a victim of the opium habit.

ALCOHOL.

16. *Alcohol* can be made out of any thing that contains sugar. It is only necessary to add yeast, or to allow it to stand uncovered in a warm place. In raising bread, the sugar which is contained in the flour is changed into alcohol and carbonic-acid gas. This is fermentation. But the heat of the oven causes the small portion of alcohol to evaporate. The carbonic-acid gas, after puffing up the loaf, and making it light, also disappears.

17. Wines are the fermented juice of grapes. They contain from five to twenty-five per cent of alcohol. Brandy, rum, whisky, and gin are distilled liquors, and are about half alcohol. Beer, ale, and porter are made from grain, and have from three to eight per cent of alcohol. Cider is made of apple-juice, and has from three to ten per cent of alcohol. The home-made wines, from currants, gooseberries, and elderberries, contain a small percentage of alcohol.

18. Wines and liquors are very commonly adulterated. The pure article is costly. By adding certain substances to an inferior wine, or even to alcohol and water, the taste and effect of good wine can be imitated, and a large profit made. Beer and porter are adulterated, to make them froth, and to improve their flavor and intoxicating qualities. Some of the substances used are cocculus indicus, alum, aloes, copperas, sulphuric acid, nux vomica or strychnine, jalap, and lime.

19. Alcoholic drinks cause a flushing of the skin and a feeling of warmth. But they do not maintain the heat of the body: they rather lessen it. This has been proved by careful experiments, and by the experience of travelers in the arctic regions.

They *check the waste and repair* that are naturally going on. This may be at times an advantage,—in a wasting disease, for example. But in health it is much better that the natural processes should be undisturbed.

20. Athletes who are training for hard trials of their bodily vigor, abstain from alcohol and tobacco. Disease, early failure of strength, and premature death, are the results of drinking-habits. Physicians, insurance companies, and all observing men, testify to this.

21. It is not always true that the strong liquors are most pernicious, nor that the milder drinks are comparatively harmless. In some sections of this country, *cider* is a worse evil than whisky. Its apparent harmlessness attracts those who would refuse stronger liquors. When the appetite for this drink is awakened, it requires for its satisfaction an amount of soaking that keeps the faculties benumbed, and reduces the individual to worthlessness as surely as more fiery, but less abundant, potations.

22. It is a melancholy fact, that "the evil that men do," in this regard, "lives after them." The iniquities of the fathers are visited upon the children. By an inflexible law of nature, the effects of alcohol are not expended upon the user alone. Morbid cravings for drink, tendencies to disease, weakness of body and mind and character, are the heritage of misery which he bestows upon his offspring.

23. **Chloral** is a drug of great value in the hands of the physician. It gives sleep to those who can not sleep in the natural way. But it is dangerous. It has caused death. A habit of using it may be acquired which is very injurious to body and mind, and very difficult to break.

QUESTIONS.

SECTION I. — 1. What is food?

2. What kinds of material are necessary for the repair of the body?

3. How many elements are found in the body? Name them. Give the number of parts in a hundred of the four most abundant.

4. Name two elements which are found in almost every article of food.

5. Do we need in our food any other elements than the fifteen contained in our bodies?

6. Where are these fifteen elements found? Name a great distinction between plants and animals.

7. What is the great work of plants?

8. Are all plants food for man? Why not?

9. What do we infer as to man's food from his teeth?

10. What article of food is the best combination of the necessary elements?

11. Of what does man's ordinary diet consist?

12. Of meats, which is the best? What objection is there to pork? What is the value of veal as food? Of lamb? Of mutton?

13. How does fish differ from flesh?

14. Which are the best of the shell-fish?

17. What are included in the term "starchy foods"? How many parts of starch in a hundred of dried wheat? Of dried oats? Of dried rice?

18. What do the grains contain besides starch? What are the advantages of the different grains?

19. What is the most popular vegetable?

20. Why are pease and beans valuable?

22. Is sugar useful as a food?

23. What is butter?

24. What is the use of cooking?

25. What is the danger in eating meat raw?

26. What gain is there in cooking starchy foods?

27. What are the essential points for good bread?

28. What makes bread *light?* What makes bread *heavy?* What makes bread *sour?*

29. Why is not pie-crust as digestible as bread? Why is food fried in fat indigestible?

31. What mineral do we take pure? Are mineral matters changed by digestion?

32. What is the proportionate amount of water in the body?

33. Is drinking-water ever absolutely pure? What sometimes happens to water which is carried through lead pipes? What kinds of pipes are safer? What are symptoms of lead-poisoning?

34. How far must a well be from all filth to be safe? Is it always safe at that distance? How may aqueduct-pipes become dangerous?

35. Can the body adapt itself to different habits of eating and drinking?

36. Do the same rules for eating and drinking always apply to all?

37. What is nature's guide to right habits of eating and drinking? Is the appetite always healthy?

38. Under what circumstances is it necessary that a large number of people should adopt the same habits of eating and drinking? What is a sufficient amount of food daily for a man?

Section II. — 1. What are stimulants? What are narcotics? Is the same thing ever both a stimulant and narcotic?

2. What service does a stimulant render in digestion?

3. What is the immediate effect of tea and coffee? Are they ever injurious?

5. Should growing children use them?

6. What is the first effect of tobacco? What are its later effects?

7. What is its effect on the appetite? on the stomach? On the nerves?

8. What is its effect on the heart?

9. Have boys any good reason for using tobacco?

11, 12. What is the effect of opium in small doses? In large doses?

13. What is the effect of the opium habit on the moral nature?

14. What is the effect of opium on the digestion?

15. Under what condition only may opium be rightly used?

16. What does alcohol come from?

17. What is wine? How much alcohol does wine contain? What

are brandy, rum, whisky, and gin? How much alcohol do they contain? What are beer, ale, and porter? How much alcohol do they contain? What is cider? How much alcohol does it contain? Do home-made wines contain alcohol?

18. How are the purchasers of wines and liquors defrauded?

19. Does alcohol keep up the heat of the body? What is its effect on the processes of waste and repair?

20. What is the general testimony as to the effects of alcohol on the health?

21. Is it true that only the stronger liquors are hurtful?

22. What is the effect of alcohol on the children of the drinker?

23. Under what condition only should chloral be used?

CHAPTER VII.

DIGESTION AND ABSORPTION.—THE LYMPHATIC SYSTEM.

SECTION I.—1. **Food** repairs the waste of the body, and keeps up life. If asked how it is made to do this, you could answer, "It is eaten." By that, you mean that it is taken into the mouth, and chewed, and swallowed. But if asked again, What becomes of it after it is swallowed, and how does it get into our bones and flesh and brains, and keep every particle living, and perhaps growing, you could not answer without study. For, in health, we know nothing about our food, by our feelings, after it is swallowed.

2. If, however, we examine into the matter, we find that the food, after being swallowed, passes on down a tube which extends through the whole length of the trunk, beginning at the lips. This tube is the alimentary canal.

3. The **alimentary canal** is about twenty-seven feet long, in a man. In order to get it into the trunk of the body, which is only about two feet long, a portion of it is coiled up in a mass. This portion we call the bowels.

4. Most of the canal is about an inch and a half wide.

SUGGESTIONS TO TEACHERS.—1. An emulsion can be made with oil and the white of an egg, or with mucilage obtained from the druggist's. This can be compared with a mixture of water and oil.

2. A tube made of chamois leather, or a cone of filter-paper, will illustrate the soaking through membranes.

In two places it spreads out, as a brook spreads into a pond. One of these enlargements is the *stomach*; the other, the *large intestine*.

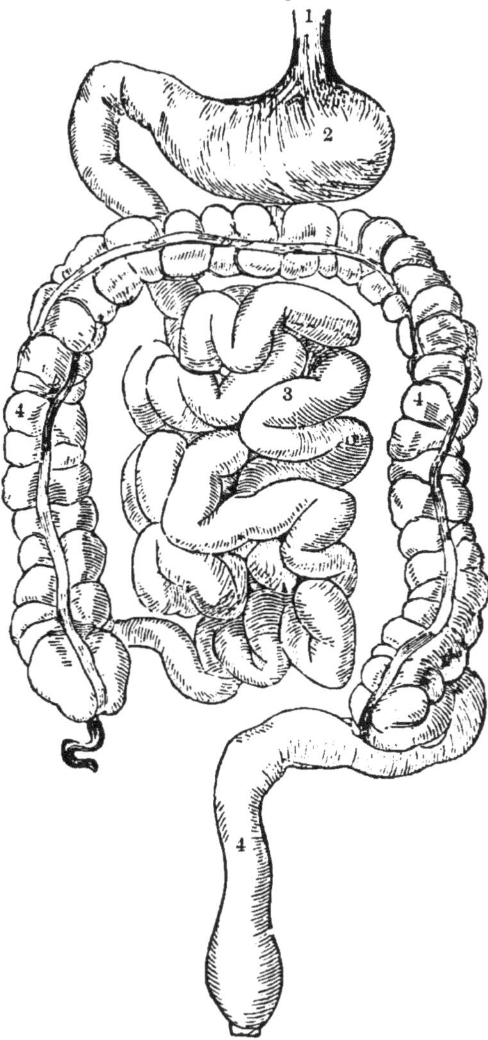

Fig. 37.

ALIMENTARY CANAL FROM THE ŒSOPHAGUS DOWN.—1. Œsophagus. 2. Stomach. 3. Small intestine. 4. Large intestine.

5. One layer of the wall of the canal is involuntary muscle. Part of the fibers run lengthwise, and part crosswise; and they surround the tube.

6. The canal is lined by *mucous membrane.* Mucous membrane can be seen on the walls of the mouth. It is quite like the skin in structure, but differs in being soft and moist and red. The mucous membrane lines the alimentary canal, and other cavities in the body, just as the skin covers the outside.

7. We have been speaking of the alimentary canal as a whole. Different portions of it have special names. The first portion is the *mouth.* Then we come to the *throat;* then to the *œsophagus,*

or *gullet;* then to the *stomach*, which lies at the lower border of the ribs, in front; then to the *small intestine*, which is twenty feet long, coiled in the lower part of the abdomen; then to the *large intestine*, which is five feet long, and ends the tube.

DIVISIONS OF THE ALIMENTARY CANAL.

1. MOUTH.
2. PHARYNX (*Throat*).
3. ŒSOPHAGUS (*Gullet*).
4. STOMACH.
5. SMALL INTESTINE.
6. LARGE INTESTINE.

8. Each of these divisions of the canal has its own peculiarities, and each has its own part of the work of the whole to do. This work is called digestion.

9. By **digestion**, we mean the changes that take place in the food, as it passes through the alimentary canal, by which it is fitted to be taken into the blood.

10. The need of such changes is very plain. We could not imagine that the food, in the condition in which it is eaten, could be taken into the blood. There are no openings from the canal into the blood-vessels. Nothing can get out of it into the rest of the body, unless it can soak through its walls, as water would soak out of the finger of a glove if it were poured into it. To do this, the food must be dissolved.

11. A lump of sugar or salt will dissolve in water; but a piece of meat, or a cooked egg, or oatmeal, or many other articles of food, will not. All such articles must be changed in the alimentary canal, so that they will dissolve in water.

12. We find, accordingly, that from the walls of the canal, and from certain organs called glands which lie just outside of the canal, and communicate with it by tubes, juices are poured out, called digestive juices, which mix with the food, and make in it the very changes that are required. These juices differ from each other, and come into different portions of the canal. One juice acts on one kind of food, and another juice on another kind.

13. The butter and cream, and all the other *fats*, that we eat, are acted on in a way peculiar to themselves. They are not really made to dissolve in water; but a juice is furnished, called the **pancreatic juice**, with which they are so thoroughly mixed, that they will pass through that part of the wall of the alimentary canal which is their special way out.

14. There are some other fluids, besides this pancreatic juice, that oil will mix with; and such mixtures are called *emulsions*. Milk is an emulsion. The oil-globules diffused through it rise to the top, and make the cream; and from the cream we make butter, by the method already described (Chap. VI., Sect. I., 23). We eat the butter. It passes down, without being changed, until it reaches the place where the pancreatic juice comes in. With that it mixes; and all these little oil-globules are separated again, and diffused through the fluid, just as they were at first in the milk. This fluid is the **chyle**; and it looks so very like milk, that the little hair-like tubes that carry it away, after it has passed through the wall of the canal, are called **lacteals**, or milk-vessels. (Latin, *lac*, milk.)

15. The *muscle* in the walls of the canal has a good deal to do with digestion, as well as the juices inside. It acts in two ways: it forces the food along, and it kneads and mixes it with the juices. We know how the muscles of the mouth and throat close about it, when we swallow. These are voluntary muscles, and we are conscious of their action. When it gets below the throat, the muscles are still contracting around it and behind it, pushing it on, as we strip water from a soft tube by drawing it between our fingers. But these are involuntary muscles, and we are not conscious of their action. While it remains in the stomach, the muscles are constantly "working" it, very much as a baker works his bread to mix the yeast with it. The same processes continue through the small and the large intestine.

If the juices are too scanty, or poor in quality, digestion does not go on well. If the muscle of the walls of the canal is weak or sluggish, digestion does not go on well. These conditions are called *dyspepsia.*

16. What has been said of the changes in the food, that constitute digestion, may be summed up as follows: *The fats* are made into a fine emulsion. *The other kinds of food* are changed into substances that easily dissolve in the fluids of the canal.

THE TEETH.

17. Let us examine more particularly the different parts of the digestive apparatus. Just behind the lips, the outer gates of the alimentary canal, stand the inner gates, the **teeth.** A child of about five years of age, who has not yet lost any of his first teeth, has twenty in all, ten in each jaw. If we could look deep into his jaw-bones, we should see, beneath these twenty, twenty-eight more,—buds of teeth, so to speak, which are his second set (all but four), almost ready to grow out. So that, at that age, a child has really forty-eight teeth,—more than at any other time in his life.

18. The *first set* are,—

Eight *incisors* (cutters), four front teeth above and below.

Four *canines* (Latin, *canis*, a dog), next to the incisors.

Eight *molars* (Latin, *mola*, a mill), grinders, back teeth.

The incisors are chisel-shaped, and intended for cutting, not for chewing.

The canines correspond to the long, pointed fangs of a dog, and are made for piercing and holding on to things.

The molars are broad and blunt, and are made for grinding up food.

19. The *second set*, the last of which do not appear before the age of seventeen years, comprises

Eight *incisors*, four front teeth above and below.

Four *canines*, next to the incisors.

Eight *bicuspids*, next to the canines.

Twelve *molars*, back teeth.

The first set come out, one by one, between the ages of five and fourteen years; and the second set appear one by one as the first are lost. The wisdom teeth are the last to appear, at some time after seventeen years of age.

20. Animals that live wholly on flesh, like the tiger, have no grinders, but only cutting-teeth. Their jaws do not move from side to side, but only up and down, like shears. On the other hand, animals that live wholly on vegetable food have many broad grinders, and not so many cutting-teeth in proportion; and their jaws have a motion from side to side, as well as up and down. Man is well provided with both kinds of teeth, which goes to prove that he was made to live on both vegetable and animal food.

21. A tooth is made of a substance like bone, called *dentine.* At the end of the root, a very small hole can be seen. This is the entrance to a canal which runs through it lengthwise, and contains the tiny nerve and blood-vessel which supply it. The root is fixed in a socket in the jaw-bone. The crown, or body, projects beyond the gums. The neck is at the junction of crown and root. The crown is covered by enamel, which is the hardest substance in the body.

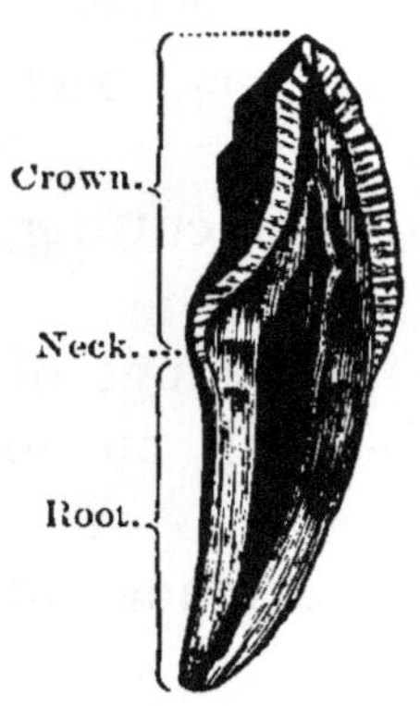

Fig. 38.
INCISOR TOOTH SAWED IN TWO.

22. It is the work of the teeth to bite off mouthfuls of food of the right size, and then to grind them up fine. This is the first act in digestion. When food is well ground, the juices of the stomach easily mix with it, and act on it. If the teeth are gone, or if food is bolted without being chewed, it enters the stomach in lumps, which the juices can not easily penetrate. When food is not soon acted on by the digestive juices, it becomes sour, and makes gases, which distend the stomach, and often give pain.

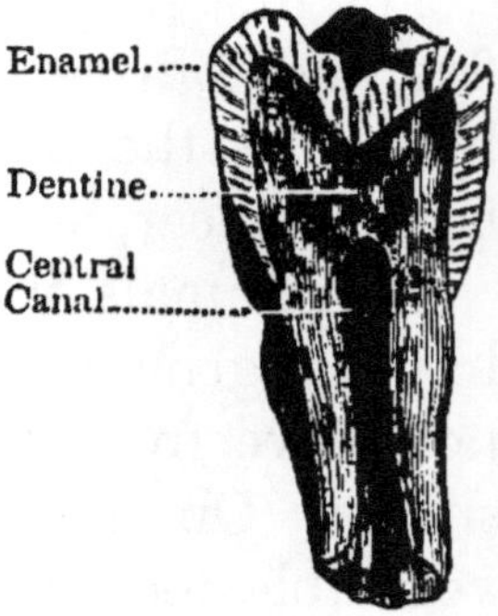

Enamel.
Dentine.
Central Canal.

Fig. 39.
MOLAR TOOTH SAWED IN TWO.

23. The teeth stop growing after they have taken their places; and, if they are chipped, they do not heal. They may be injured,—

1. *By very hot or very cold* substances.
2. *By some medicines.*
3. *By decomposing food* between them.

They should be *brushed every day,* and cared for by a dentist if unsound. Good health and good looks both depend much on them.

THE SALIVARY GLANDS.

24. In health, the mouth is always moist. When we taste any good thing, or even think of it, the "mouth waters." This water is the saliva, or spittle; and most of it comes from bodies called the salivary glands.

25. There are three pairs of these glands, — two parotid glands, two submaxillary glands, and two sublingual glands.

Fig. 40.

SALIVARY GLANDS. — 1. Parotid gland. 2. Submaxillary gland. 3. Sublingual gland.

The *parotid gland* is the largest, and is situated just under the ear. Mumps is an inflammation of this gland, which makes it swell, and the neighboring parts of the neck with it.

The *submaxillary gland* is next in size, and lies behind the edge of the under jaw beneath the floor of the mouth.

The *sublingual gland* lies farther forward than the submaxillary, also under the floor of the mouth.

26. Each one of these glands has one or more tubes, called ducts, opening into the mouth. The duct of the parotid gland opens on the inside of the cheek. The ducts of the submaxillary and sublingual glands open under the tongue.

27. These glands make the **saliva** from the blood which passes through them. The saliva moistens and softens the mouthful of food, coats it over, so that it will slip down the canal, and to some extent dissolves it. When the mouth is absolutely dry, it is almost impossible to chew or swallow any thing.

THE STOMACH.

28. The stomach is a portion of the alimentary canal about twelve inches long, expanded into a pouch. It lies at the lower border of the ribs in front, more on the left side than on the right. When it is empty, it collapses, like any other bag. When it is full, it extends down below the ribs; and, as it lies directly under the heart, it sometimes presses up, and makes the heart feel crowded.

29. This pouch, like the rest of the alimentary canal, is lined with mucous membrane, somewhat like the lining of the mouth. If this membrane is examined with a lens, innumerable little holes appear: so small, and so close together, are they, that it has been estimated that there are five millions of them in all. These are the mouths of little pits that dip down in the membrane. Some of these pits are shaped like the finger of a glove: others have side-branches, like several glove-fingers opening into one central finger. These are the glands of the stomach, which make the **gastric juice.** When food enters the stomach, this juice wells up from the pits until a drop stands at the mouth of each one. This overflows, and another follows; and so it keeps coming, until there is just enough to mix with the food taken. Then it stops. Meantime the muscular walls have been turning and squeezing the food, somewhat as the teeth and tongue do in the mouth.

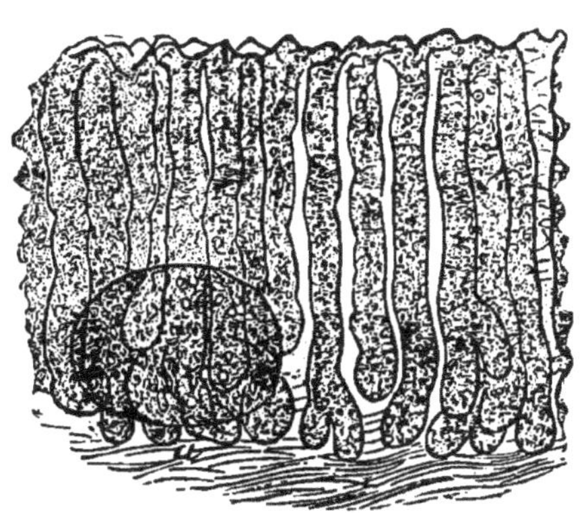

Fig. 41.
STOMACH GLANDS OF A PIG.

30. The outlet of the stomach is at its right end, and is guarded by the **pylorus** (from a Greek word, meaning keeper of the gate). This pylorus is a ring of muscular fibers, which surrounds the canal, and, by contracting, closes it. Its duty is, to let no food pass out until it has been properly acted on by the stomach. When the food is hard and indigestible, the laboring stomach often becomes exhausted and distressed. It would fain get rid of its contents: but the pylorus steadfastly resists, until vomiting occurs; or else, this resistance being overcome, the troublesome matters pass down, to cause similar discomfort in the intestines.

31. Near the beginning of the small intestine, two little tubes open into it by the same orifice. If we follow these back a short distance, one will lead us to the liver, the other to the pancreas.

THE LIVER.

32. The **liver** is a large organ situated at the lower border of the ribs, on the right side. It is a gland, and it does three things:—

1. It helps to *purify the blood,* by taking out of it certain substances.

2. It *makes glycogen.* Glycogen is a food substance, made of the food we eat, and stored up for a time in the liver,

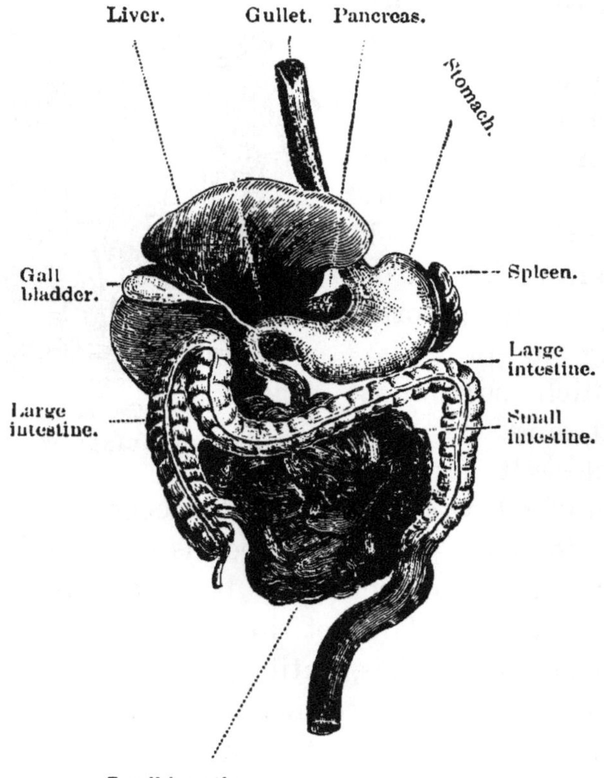

Fig. 42.
THE LIVER AND OTHER ORGANS OF DIGESTION.

somewhat as starch is stored up for the use of a potato-plant in the potato.

3. It *makes bile.*

33. So large an organ as the liver must have a very important work. When the flow of bile is cut off, the

flesh is rapidly lost, and death ensues. Sometimes the duct of the liver gets clogged. The yellow stream of bile is dammed up, and forces itself into the blood-vessels, and is carried all through the body. It colors the skin and eyes yellow. This condition is called *jaundice.*

The bile in the intestine mingles with the other digestive juices, and with the food, and aids digestion in several ways.

34. The **pancreas** lies across the backbone, just behind the stomach. It is only one-twentieth as large as the liver. It is a mass of little tubes, in which the *pancreatic juice* is forming during digestion. These little tubes empty into one large tube, which runs through its whole length, and finally discharges into the intestine, just as the small drain-pipes in the houses of a town empty into the large main in the street, and that, finally, into the river.

35. The pancreatic juice makes a mixture, called an emulsion, of the fats, in which they can easily pass through the walls of the canal. It also aids in dissolving other parts of the food.

36. The *lining* of the whole length of the small and large intestines contains little pits similar to those which are found in the stomach. A fluid called the *intestinal juice,* which helps digestion, wells up out of them; but the action of this fluid is not so important as that of the gastric juice.

37. **Recapitulation.** The digestive apparatus consists of the alimentary canal, and certain glands connected with it.

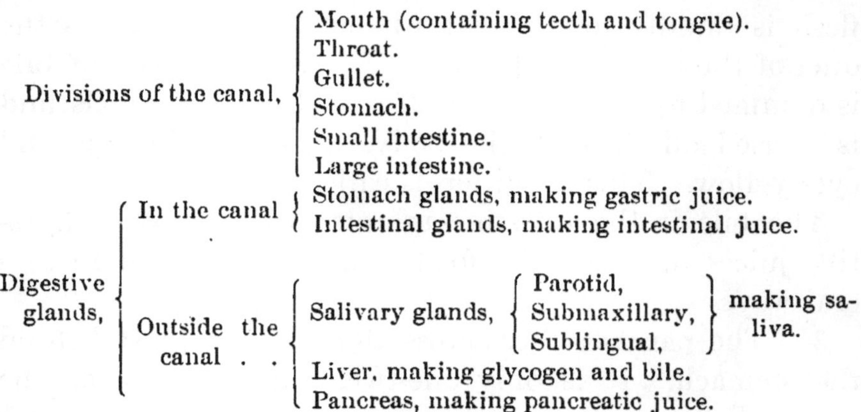

- Divisions of the canal,
  - Mouth (containing teeth and tongue).
  - Throat.
  - Gullet.
  - Stomach.
  - Small intestine.
  - Large intestine.
- Digestive glands,
  - In the canal
    - Stomach glands, making gastric juice.
    - Intestinal glands, making intestinal juice.
  - Outside the canal . .
    - Salivary glands, Parotid, Submaxillary, Sublingual, making saliva.
    - Liver, making glycogen and bile.
    - Pancreas, making pancreatic juice.

ABSORPTION.

SECTION II.—**1.** The **alimentary canal** may be likened to the kitchen in which the food is prepared. The **blood** is the carrier, swiftly moving through the passage-ways, and serving every room in the house of life.

2. Through the whole length of the canal, its wall is filled with meshes of hair-like blood-vessels, completely surrounding it, like a net-work, ready to soak up the food through their thin walls just as fast as it is made ready. Through these little vessels the current is moving, so that fresh blood is continually taking the place of that which has just passed on with its load.

3. Let us follow *a mouthful of food*, consisting of bread and meat, from its entrance between the lips into the canal. First the teeth cut and grind it, the tongue and cheeks skillfully turning and shifting it from side to side, the saliva, meantime, wetting and partly dissolving it. When it is fine and soft enough, the tongue forces it against the roof of the mouth, and so slips it back to the throat. Certain muscles then contract, and lift the throat up around it. It is clasped and pushed down by the contraction of successive rings of muscular fibers, through the gullet into the stomach. There it remains for a time. It has not been long in the stomach before portions of it are digested, and these begin to pass through its walls. Other portions are not digested by the gastric juice. But the whole is reduced to a fluid-like gruel, and then passes the pylorus into the small intestine. In this state it is called **chyme**. Here it meets the bile and the pancreatic juice. These mix with the fat, and make an emulsion which is called **chyle.** The rest of the food is also changing, and being dissolved out of the branny and fibrous parts which can not be digested.

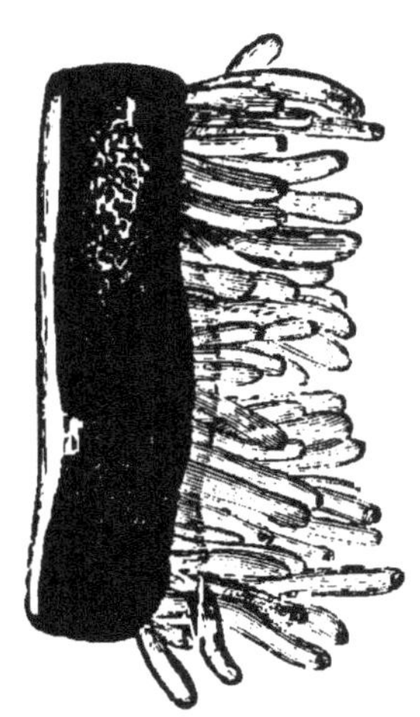

Fig. 43.
VILLI OF THE SMALL INTESTINE MAGNIFIED.

4. In the lining of the small intestine, we find a special apparatus for absorption; and it is here that absorption goes on most actively.

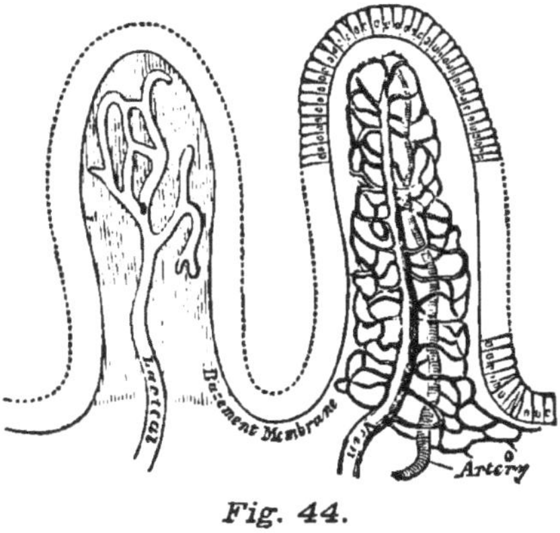

Fig. 44.
SHOWING THE VESSELS IN VILLI.

This lining has a soft look, like velvet. With a microscope we can see that the reason that it looks so, is, that it actually has a nap, like velvet. This nap is made of innumerable short, thread-like projections, which are called **villi** (Latin, *villus*, a tuft of hair).

5. Each *villus* contains a net-work of fine blood-vessels, and, also, one of another

kind of vessels called lacteals, which will be described presently, whose special work it is to take up fat.

6. As the muscular walls of the intestine contract and relax, these little villi are worked in the mass of food; and they draw in the digested part as the fine rootlets of a plant draw up liquid nourishment from the earth in which they stand.

7. As the chyme and chyle pass on down the small intestine, the digestible portion is constantly growing less, until at length it has all been taken up,—the chyle chiefly by the lacteals, and the other portions of the food chiefly by the blood-vessels; and what remains is indigestible and useless.

8. The food, having thus become a part of the blood, is carried through the body, and permitted to soak out through the walls of the capillary vessels, to feed each particle of living substance.

9. Water, and mineral matters like salt, that are dissolved in water, need no digestion, and are taken up by the vessels in all parts of the canal.

SECTION III.—**The lymphatic system.** 1. This is a system of tubes and glands,—the tubes resembling the blood-vessels in some respects. They begin with hair-like tubes running among the capillaries, and much like them. These unite to form larger tubes, which unite with others, and so on, until they have all been united into two tubes, each about as large as a slate-pencil. These open into the large veins, not far from the heart. They are called the **thoracic duct** and the **right lymphatic duct.**

2. But the lymphatic system is not just like the system of blood-vessels. The **lymph,** as the fluid which they con-

tain is called, does not "circulate." The blood starts from the heart, and is brought around back to the heart again. The lymph starts from all parts of the body, and is

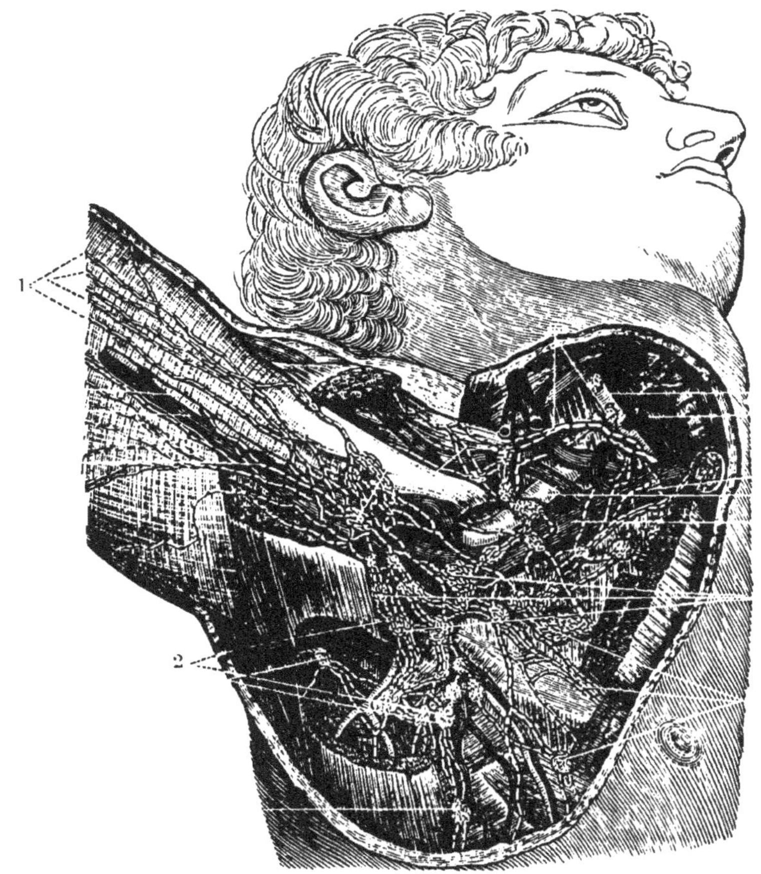

Fig. 45.

LYMPHATICS. — 1. Lymphatic ducts. 2. Lymphatic glands.

brought in toward the heart. If you compare the lymphatic vessels with the capillaries and veins, you have nothing in this system to correspond with the arteries.

The lymphatic vessels are much more delicate and slender than the blood-vessels.

3. When a farmer has a wet field, he frequently lays pipes in it made of burned clay, called tile. The standing water soaks into these pipes, and is carried off; and so the field is dried.

The lymphatics are the drain-pipes of the body. They assist the blood-vessels in taking up the fluids which are standing in all parts, and carrying them away to be delivered up to the blood-stream at the proper place.

4. The **lacteals** comprise that portion of the lymphatic vessels that begins in the walls of the small intestine. When digestion is not going on, they are drain-pipes, like the rest. As soon as digestion begins, they begin to look white and milky. They are then engaged in their special work of taking up the fat from the intestine. They are found in all the villi.

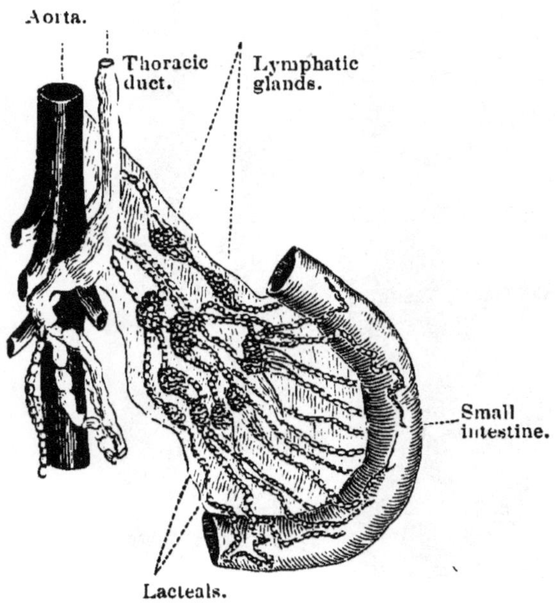

Fig. 46.

LYMPHATICS OF THE INTESTINE.

SECTION IV.—1. The **process of digestion** is not under control of the will. In health it takes care of itself, and we give no thought to it. But, when the digestive apparatus is weak or disordered, it becomes the seat of almost constant pain and discomfort. The appetite is lost, the strength fails, the nerves become irritable, and the brain is clouded. It is important, therefore, to know what helps or hinders it.

HINDERANCES TO DIGESTION.

1. *Eating too fast.* In this case, the food is taken down without being prepared by chewing.

2. *Strong excitement.* Sudden fear, anger, or grief takes away the appetite, and stops the flow of the digestive juices.

3. *Great fatigue.* No careful horse-owner will feed his animal immediately when he comes in tired and heated. Food swallowed under such circumstances will be digested with difficulty by a man or a horse.

4. *Mental effort.* It is not well to read, or to study during a meal. The mind should be at rest, and some degree of attention should be given to the food.

5. *Too much food.* Evidently there is a limit to the amount of food that can be digested. The stomach may be so full that it can not easily move its contents. The quantity may be so great that the digestive fluids can not fully permeate them. Those parts of the mass which are not digested will soon decompose, producing acidity, and a pressure of gas.

6. *Too much liquid with food.* A good deal of water is directly absorbed. But, when too much is taken, some remains in the stomach, and so dilutes the gastric juice, that it is weakened.

7. *Very cold substances,* as ice-water, taken with food, will sometimes stop digestion. The gastric juice acts best in a

temperature of about a hundred degrees. Stomachs, however, differ in their ability to withstand cold as much as the outer parts of the body.

8. Irregularity in eating. The digestive apparatus is subject to habit, like the rest of the system. At the accustomed meal-time, the saliva and the other digestive fluids will flow, though no food is taken. When that time is passed, they do not start so readily.

9. Lack of exercise. In a sluggish condition of the body, the digestive juices flow slowly. The alimentary canal does not contract vigorously to knead the food.

10. In applying these principles, it is to be remembered that one person can do without harm, and sometimes with advantage, what is injurious to another. Men differ very widely in their habits. Each should understand the facts and principles of physiology, and apply them, with the aid of experience, to his own case.

EFFECTS OF ALCOHOL ON DIGESTION.

SECTION V.—**1.** Alcohol irritates the mucous membranes. You could not hold it in your mouth; and, if you should swallow any of it clear, your stomach would seem burned. When mixed with much water, as it is in wines and liquors, it is less fiery, and the sensations produced in the mouth and stomach may be agreeable.

2. It is by this **irritant quality** that it injures the stomach. In the famous case of St. Martin, the inside of whose stomach could be observed through a wound, Dr. Beaumont found that "the free use of ardent spirits, wine, beer, or any of the intoxicating liquors, when continued for some days," constantly produced "a state of inflammation and ulceration in the lining membrane, and change of the gastric juice." At the same time, St. Martin suffered no pain, or other sensations, which indicated the true state of things.

Examinations of the stomachs of drinkers, after death, show the same conditions.

3. One of the most wonderful things about our bodies is, that they will *change so as to suit the conditions* in which they are. If a man has to work with his arms, they grow large and strong. If the skin is exposed to the weather, or to chafing, it loses its smoothness, and becomes hard. The stomach was made to digest wholesome food. If a man insists on making it a receptacle for burning liquors, Nature straightway begins to adapt it to that use. The delicate membrane *grows tough*. The mucus which naturally moistens it becomes thick and ropy, so as to protect the surface. Some of the little glands which pour out the gastric juice are destroyed. Those which remain are unhealthy. If the process is carried far enough, he has a pouch which gladly receives alcohol, but its usefulness as a stomach is greatly impaired.

4. If we put in a test-tube some *gastric juice* from the stomach of an animal, and add alcohol to it, a white powder appears, and settles to the bottom. This is **pepsin**, which gives the juice its power to digest. Alcohol separates it from the juice; so the drinker, with a stomach perhaps already weak, is adding to his gastric juice that which destroys, for the time, what activity it has. After the alcohol passes out of the stomach, the pepsin is dissolved again.

5. The effect of alcohol on the intestine is similar to

that on the stomach, but much less; because most of it is absorbed before reaching that part of the canal.

6. Alcohol is carried to the *liver* by the blood.

Alcohol irritates the liver, and causes an overfullness of its blood-vessels.

Alcohol often excites inflammation of the liver. This is followed by various changes, one of the most notable of which is a contraction into a hard, knobbed mass, called, in medical works, *gin-drinker's liver*.

7. Thus the power of digesting food, which is the source of health and strength, is assailed by alcohol at two important points, — in the stomach, and in the liver.

8. The effect of alcohol on the **kidneys** may be referred to here, because it is similar to its effect on the liver. The kidneys are purifying organs. The blood is constantly passing through them, and they filter out of it waste matters. Blood charged with alcohol irritates them, and excites, frequently, a slow inflammation, which results in their destruction. This is one of the forms of **Bright's disease**, and in most cases is incurable.

9. These effects do not always follow the use of alcohol, but in very many cases they do. They are the possibilities and dangers of the drug.

QUESTIONS.

SECTION I. — 1. Why do we need food?
2. What is the alimentary canal?
3, 4. What is its length and diameter?
5. What important tissue in its wall?
6. What is mucous membrane?
7. Give the divisions of the alimentary canal.

9. What is digestion?

10. Why is digestion necessary?

11. Is all our food easily dissolved?

13, 14. How are fats digested? What is an emulsion? What are the lacteals? What is chyle?

15. What has the muscle of the alimentary canal to do with digestion? Is it voluntary, or involuntary?

16. What are the changes that constitute digestion?

17. How many sets of teeth do we have? When do we have the largest number of teeth in our jaws?

18. Give the number and names of the first set.

19. Give the number and names of the second set.

20. How do the teeth indicate the habits of the animal?

21. What is the structure of a tooth?

22. What is the work of the teeth?

23. How may the teeth be injured?

24. How is the mouth kept moist?

25. Name the salivary glands. Give their location.

26. What are their outlets?

27. What is the use of saliva?

28. Where is the stomach situated?

29. Describe the glands of the stomach. How is the gastric juice mixed with the food?

30. What is the pylorus?

31. What two large glands are connected with the upper end of the small intestine?

32. Where is the liver situated? What three things does it do?

33. What is jaundice? What does the bile do?

34. Where is the pancreas? What is its structure?

35. What does the pancreatic juice do?

36. What glands are there in the walls of the small and the large intestine? What is the fluid that comes from them called, and what is its use?

37. Give the divisions of the alimentary canal. Enumerate the digestive glands in the canal. What is their product? Enumerate the digestive glands outside of the canal. What is their product?

SECTION II.—1, 2. How is digested food carried through the body?

3. Follow a mouthful of food in its progress through the alimentary canal.

4. What special apparatus for absorption in the small intestine?

5, 6, 7. Describe its action.

9. Is water digested?

SECTION III.—1. What is the lymphatic system? What are the names of the tubes in which the lymphatic vessels terminate?

2. State the difference between the lymphatic system and the circulating system.

3. What is the use of the lymphatics?

4. What are the lacteals?

SECTION IV.—Name some hinderances to digestion.

SECTION V.—1. What is the effect of alcohol on the mucous membrane?

2. What was its effect on the stomach of St. Martin?

3. How does the stomach adapt itself to alcohol?

4. What effect has alcohol on the gastric juice?

5. What effect on the intestine?

6. How does alcohol reach the liver? and how does it sometimes affect it?

8. What is the effect of alcohol on the kidneys?

CHAPTER VIII.

RESPIRATION AND THE VOICE.

SECTION I.—**Air.** **1.** We have studied the processes of digestion and absorption, by which food and water get into the blood, to be distributed through the body. It remains to study the method by which air is taken in and used. The process is called respiration.

2. We live in air, and we can not live out of it any more than fishes can live out of water. We can not see the air, but we can feel it. When it is moving very rapidly, it has great force. It can root up trees, and carry away houses. If we could see it, it would appear like water. When the wind is blowing, we would see a stream of it pouring across the country like a river, or like a flowing sea.

SUGGESTIONS TO TEACHERS.—SECTION I. A few chemical experiments will illustrate this section. Show the effect of oxygen and of carbonic-acid gas on a burning candle. Fill a test-tube with lime-water, and breathe into it to demonstrate carbonic-acid gas by the formation of a milky precipitate of carbonate of lime.

SECTION II. Show, if possible, a fish's gills. Illustrate by drawings; by a bunch of grapes, etc. Get a piece of lung from the butcher's, or, still better, the lungs and air-passages of some small animal.

SECTION III. The nasal cavities should be shown, either in the human skull, or in that of a sheep or other animal.

SECTION IV. Simple experiments with the apparatus mentioned in the text, will illustrate the principles.

SECTION VI. Show the moisture in the breath by breathing on glass.

SECTION VII. A larynx can easily be got from a butcher, and the glottis and vocal cords and muscles can be shown.

3. **Air** is a material substance, though invisible. It is in the gaseous condition. It is composed chiefly of three gases mixed.

COMPOSITION OF AIR.

Nitrogen, about . .	. 79	parts in a hundred.
Oxygen, about	21	" " "
Carbonic-acid gas, about	.04	" " "

Besides, there is watery vapor, and a little of various other substances.

4. *Nitrogen gas,* which constitutes nearly four-fifths of the air, is not used either by animals or plants. Neither is it injurious. Its office is to dilute the oxygen. If we put a lighted candle into a jar containing pure oxygen, it will burn up very rapidly. So, if an animal is made to breathe pure oxygen, it is greatly excited; and its life, which is partly a burning, is quickly burned out. The nitrogen in the air dilutes the oxygen to just the right strength.

5. *Carbonic-acid gas* is not used by animals. On the other hand, they are constantly throwing it off as a waste product. But plants live on it: it is a large part of their food. As they take it in, they give off oxygen.

Animals take in *oxygen*, and give out *carbonic-acid gas.*

Plants take in *carbonic-acid gas,* and give out *oxygen.*

Fires, and the decay of animal and vegetable matters, are adding constantly to the carbonic-acid gas. But there is only one process which is adding to the oxygen; and that is, the action of the leaves of plants. On the whole, the addition and subtraction are equal. The amount of carbonic-acid gas in the whole air remains the same, though in a smoky city it is greater than it is in the country.

Fires Decaying matters Living animals	}	. add to the *carbonic-acid gas.*
Living plants	. . . . . .	. add to the *oxygen.*

6. *Oxygen* is the most remarkable of the elements. Three-fourths of the material of which our bodies are made is oxygen: eight-ninths of the material of which water is made is oxygen. It is more than half the weight of many other substances. We can not live five minutes without a fresh supply of it. It is infinitely more valuable than gold, but is free to all. We do not have to work for it, as we do for food: we take it from the air we breathe.

THE LUNGS.

SECTION II.—1. The whole of the air that we take in does not enter the blood. Just as the digestible parts of our food are absorbed, while the indigestible parts are cast out, so that part of the air that we need, which is oxygen, is absorbed in the lungs, and the rest is breathed out.

2. The lungs are masses of little cells, with very thin walls, placed in the chest. On the outside of the walls of these sacs are the capillaries,—millions of them. They cover the sacs as the netting covers a balloon, only they are much closer than such a net. The blood from the pulmonary artery pours into these capillaries, and so spreads out all over the lung-cells. The oxygen, then, has only to pass through the thin lung-cell and the thin capillary wall to enter the blood.

3. Frogs get a part of their oxygen through their skins. These are delicate and moist, and just beneath them the net-work of capillaries is spread out. Oxygen easily

passes through. As there is considerable air dissolved in the water, they can live under it all winter on what oxygen they can take in from it through their skins. Man's

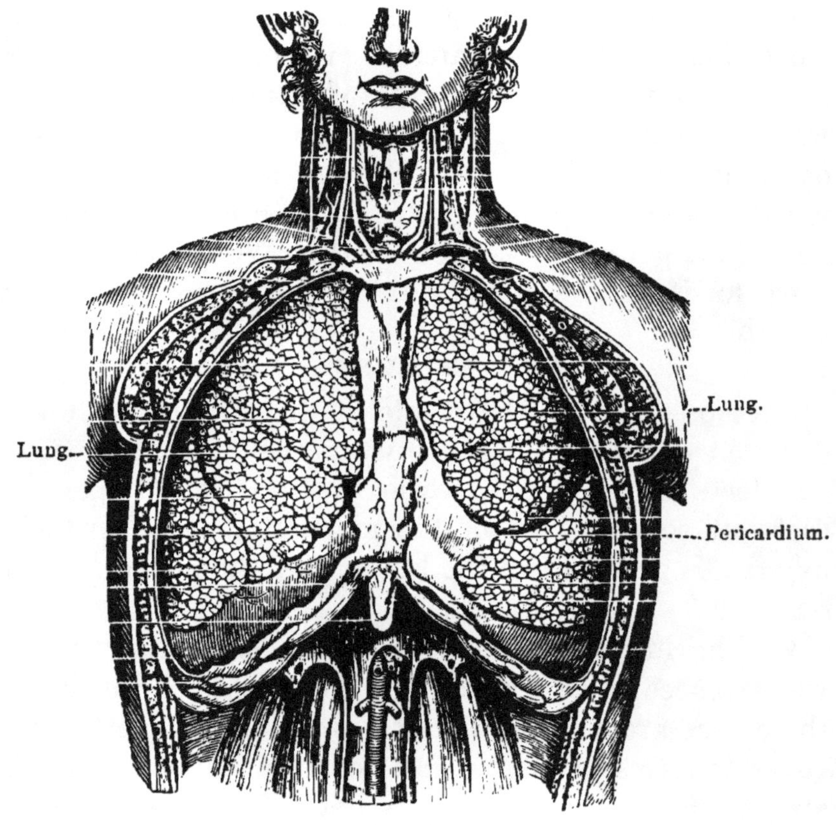

Fig. 47.
LUNGS IN POSITION, THE WALL OF THE CHEST BEING CUT AWAY.

skin is thicker, and does not allow the oxygen to pass through.

4. Fishes have a peculiar way of getting their oxygen. They, of course, must take it from the air which is dissolved in the water. On each side of their throats they

have an opening, in which are a number of flat plates of membranes, called gills, with net-works of blood-vessels in them. The water is swallowed, and passes out between the plates through these openings. The oxygen which it contains enters through the thin membranes into the vessels.

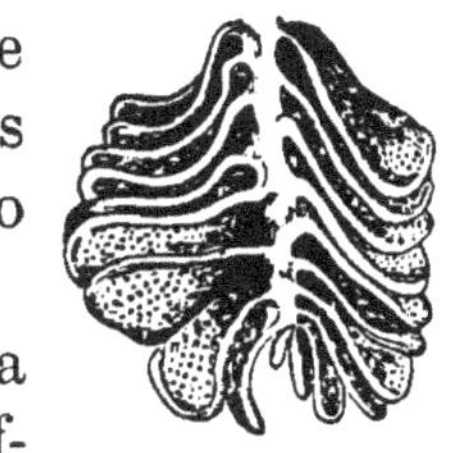

Fig. 48.
GILLS OF AN EEL.

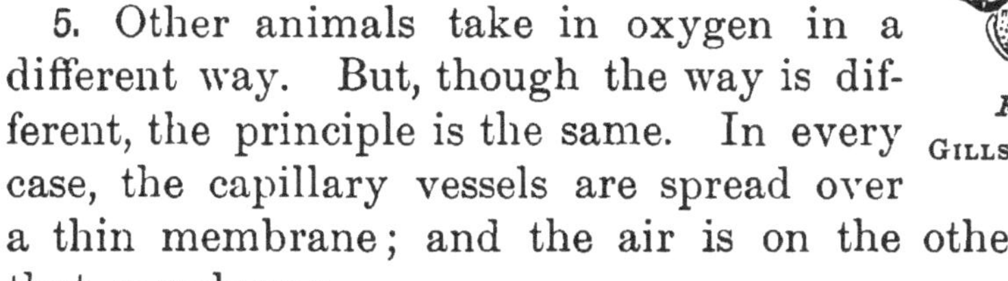

5. Other animals take in oxygen in a different way. But, though the way is different, the principle is the same. In every case, the capillary vessels are spread over a thin membrane; and the air is on the other side of that membrane.

6. In some of the lower animals, the lung is a single sac, like a bladder, with the capillaries on the outside. In man, instead of being a single sac, each lung is a vast number of very small sacs bound together in a mass, with fine blood-vessels and air-tubes surrounding and connecting them.

Imagine a bunch of grapes with the contents of each grape taken out, leaving only the skins to represent the air-cells. Suppose the stems to be hollow, and they will represent the air-tubes. Now, put several such bunches together, so that the main stems all join in one large stem, and you have something which represents the air-cells and air-tubes of a lung. To make it complete, suppose all the grapes to be joined by fine threads, like a spider's web. This represents the fibrous tissue, which is quite elastic. Suppose a blue tube to run along the main stem, and divide every time the stem does, until its small branches finally reach every grape, and form a net-work on it. From that net-work suppose red tubes to run back by the sides of the blue ones, joining, constantly, other tubes, until all are united in two large ones on the main stem of our bunches of grapes. The blue tubes are the pulmonary artery and its branches. The net-works on the grapes are the capillaries, and the red tubes are the pulmonary veins.

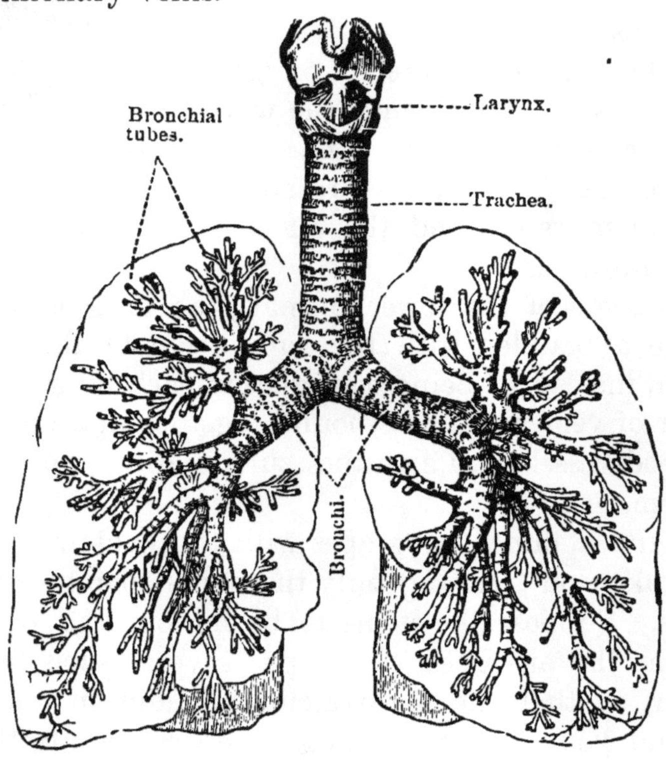

Fig. 49.

SECTION OF THE LUNGS, PARTLY SHOWING THE COURSE OF THE BRONCHIAL TUBES.

In such a figure we have the air-tubes and air-cells: we have the blood-vessels, and we have the fibrous tissue. If we add the lymphatic vessels and nerves, we have all that makes up the lung.

7. A tree is another illustration of a lung. The trunk represents the bronchus, or great air-tube, which enters the lung. The branches represent the small (bronchial) tubes. The leaves represent the air-cells. If we suppose tubes to be laid along the trunk and branches, constantly dividing, until finally each leaf is covered with a network from which other tubes come off, and run back toward the trunk, uniting as they go, we shall have again something like a lung, with the fibrous tissue, the nerves, and lymphatic vessels and glands, still to be added.

8. Do not suppose, however, that, on examining a lung, you would see all these tubes and cells. On the contrary, you would see only a light-gray or pinkish substance, mottled with black spots, very smooth on the outside. If you were to cut into it, it would look a little like fat. The cells, and many of the tubes, are so small, that only very careful study with the microscope has given us our knowledge of them.

THE AIR-PASSAGES.

SECTION III.—1. The passages by which the air gets to the lungs, are,—

The nose,	The larynx, } (windpipe).
The mouth,	The trachea, } (windpipe).
The throat,	The bronchi.

2. The **nose** is the true breathing passage. It consists of two parts:—

1. The triangular projection from the face. This is partly cartilage, which is flexible, and partly bone.

2. The cavities behind, called the *nasal fossæ.* There are two of these, corresponding to the two nostrils, separated by a partition. They are narrow, but quite extensive. They enter into the upper part of the throat by two openings, like the two on the face. Above, they connect with cavities in the forehead, and at the sides with cavities in the upper jaw-bones, and with the eye. These passages and cavities are lined with mucous membrane; and when this is inflamed, as in a cold in the head, we have a feeling of fullness, not only in the nose, but also in the forehead and eyes.

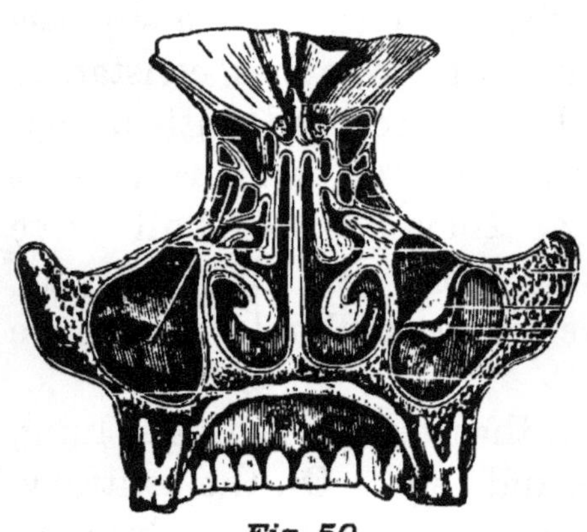

Fig. 50.
SECTION OF NASAL FOSSÆ, SEEN FROM BEHIND.

3. The *nerves of smell* are up in the top of the nasal cavities. In ordinary breathing, the air passes straight back to the throat; but, when we wish to smell any thing distinctly, we sniff suddenly, and so draw the air up to the top of the cavities, with the odorous particles in it.

4. The odorous particles will diffuse themselves through the cavities, and reach the nerves, without this sniffing, unless they are few and faint. But, by sniffing the air, we get the sensation more quickly and keenly.

5. A horse can not breathe through his mouth, but a man can. We do so when the nose is stopped, and when we are breathing rapidly, as in exercise. Some always do so when asleep.

It is better to breathe through the nose, because, —

1. The nasal cavities being narrow, the air is spread out

in a thin sheet, and so is warmed by the warm walls of the cavitics. It passes through the mouth in a large stream, and pours into the lungs without being properly warmed in cold weather.

2. Dry air, passing through the nasal cavities, is not only warmed, but moistened. When it is taken through the mouth, it dries the throat, as any one can learn by trying it rapidly for a few moments. Hence, the practice tends to produce sore-throat.

3. *Snoring* is a result of sleeping with the mouth open. The soft palate, which hangs like a curtain between the passages from the nose and the mouth, is relaxed in sleep. The two currents of air, one on each side of it, cause it to vibrate rapidly, just as a sheet will flap and rattle in the wind.

6. The habit of breathing through the nose, both in waking and sleeping hours, should, therefore, be cultivated. One of the best safeguards against catching cold in the throat, on going from a heated room into the night air, is to keep the mouth shut.

7. The throat is both a food-passage and an air-passage. Air is sometimes swallowed into the stomach: food and drink sometimes enter the wind-pipe; but, if they do, there is a great disturbance, and they must be coughed out at once. Commonly, each takes its proper course,—the food through the gullet, the air through the wind-pipe, and that without any thought on our part. The apparatus is self-regulating. When we swallow, the breathing-passage closes up tight; and a kind of trap-door, called the epiglottis, shuts down over it. You can not swallow and breathe at the same time.

8. The **larynx**, which is the upper part of the wind-

pipe, is a kind of a box made of cartilage. It is wider than the rest of the wind-pipe, and projects in front, making Adam's apple. It is the voice-box. In it are the *vocal cords*, or bands, by which sound is made. It will be more fully described hereafter.

9. The **trachea** is a tube four or five inches long, and from half an inch to an inch in diameter, extending from the larynx down into the chest. It is made of a series of flat rings of cartilage, sixteen or twenty in number, which do not quite come together behind, but are somewhat like a horseshoe. These are connected, and covered with membrane and muscular fibers. They serve to keep the tube open.

10. The trachea divides into two branches, called **bronchi.** One goes to the right, and one to the left, lung. They are made just like the trachea, and are one or two inches long.

11. Each bronchus, as it enters the lung, divides into smaller tubes, called **bronchial tubes**; and these keep on dividing until they are not more than $\frac{1}{100}$ of an inch in diameter. They then end in the air-cells.

BREATHING.

Section IV. — 1. We have studied the lungs, and the passages leading to them. We have now to study the process by which air goes in and out of them. It is not enough that the passage is open. The lungs are not like a house, with windows and doors, through which the breeze plays freely. They are, rather, like a deep well, or a mine, into which fresh air will not go, unless, in some way, a current is made. We make this current by breathing.

2. If you watch the breathing, you will see two regular movements. First, the chest and abdomen seem to swell; and then, in a moment, they fall back to their former size. This is repeated with every breath. The chest does enlarge; and, as it enlarges, the air rushes in, to fill the extra space. As it grows smaller, the same amount of air that entered is squeezed out again.

Why does the air rush in, when the space is made larger? Because air is like water: it is pressing in every direction. If you put water in a tub with a pipe opening out of it, it will force itself into the pipe. If you put a bottle under water, with the mouth up, the water will push in, and fill it. We are in the air as the bottle is in the water. If the bottle is made of rubber, the water will flow out when we squeeze it, and will flow in again when it expands. The chest may be likened to such a rubber bottle.

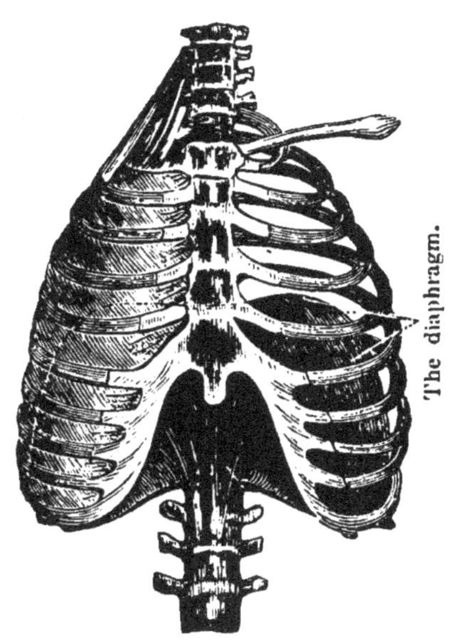

Fig. 51.
THE CHEST.

The chest is like a bellows, with one exception: in a bellows, the air enters by one opening, and goes out of another; in the chest, air enters and goes out by the same opening.

3. How is the chest made larger? In two ways,—

1. By the descent of the diaphragm.

2. By the raising of the ribs and breast-bone.

The diaphragm is attached to the lower edge of the walls of the chest, and stretches across, separating chest from abdomen, forming the floor of one and the roof of the other. But it is not a flat floor. It arches up into the chest, as may be seen in the figure, and is supported in this position by the contents of the abdomen beneath it. When we take a breath, the diaphragm contracts,—being partly muscle,—and the top of the arch is flattened, making more room in the chest. At the same time, the contents of the abdomen are pressed down, and the walls of that cavity expand to accommodate them. The abdomen is not made larger when we take breath. It expands in front just enough to make up for what it loses by the flattening of its roof.

4. But the chest is enlarged, not only by the descent of its floor, but also by the raising of the ribs and breast-bone. The ribs are joined to the backbone behind, and connected with the breast-bone by the costal cartilages, which will bend. To the ribs are attached muscles which pull them up, and others which pull them down. The former are called *inspiratory* muscles, and the latter *expiratory* muscles. Now, clasp your hands, and extend your arms in front of you, slanting a little downward. Your hands represent a portion of the breast-bone, and your arms two of the ribs. Keeping your hands clasped, raise your elbows as much as you can, and raise your hands a little. That is the motion that the inspiratory muscles give to the ribs and breast-bone. You can see that the space included between them is enlarged in both directions.

5. We have been speaking of the cavity of the chest. The air enters the lungs. But the lungs occupy almost

all of the chest, except the space filled by the heart. They are attached to its walls and floor. They are elastic; and so, when the chest enlarges, the lungs enlarge with it, and the air-cells open wide. When the chest sinks back to its former size, the elastic lung shrinks too; and the air is driven out.

6. The lungs are held to the chest-wall by the **pleuræ**. These are two empty, air-tight sacs, one for each lung. One layer covers the lung; and the other lines the chest-wall, and is attached to it. The insides of these two sacs are moistened by a fluid, so that there is no friction between the lung and the chest: and, as the sacs are air-tight, the layers can not be separated; and the chest-wall, in its movements, pulls the lung with it, just as the circle of wet leather, which the boys call a sucker, lifts a stone. The *pleura* is like the inner portion of the pericardium.

7. To recapitulate briefly. The lungs, consisting largely of little air-cells, with the air-tubes leading to them, are elastic. They nearly fill the chest, and cling closely to its walls and floor. When the chest expands by the descent of its floor and the ascent of its walls, the lungs expand with it. The air-cells open, and the outside air is pushed in to fill them,—as the air enters a bellows when we separate its walls. When the inspiratory muscles stop pulling, the chest settles back to its former size, the expiratory muscles helping it a little, and sometimes a good deal.

8. We breathe without thinking of it in sleep as well as when awake. We can not refrain from it for many seconds. Constant breathing is necessary to life, so it must be independent of our wills.

SECTION V.—1. Part of the oxygen of each breath taken in, after reaching the air-cells, goes through their walls, and through the walls of the capillaries outside of them, into the blood. It joins itself to the red corpuscles; and they float on with it through the heart, and out into the aorta, and finally into the capillaries of all parts of the body: there it leaves the corpuscles, drawn by a more powerful attraction through the capillary walls, to help to nourish and build up the surrounding substance.

2. The blood, as it discharges its load of oxygen from the capillaries, takes up, at the same place, a new load of *carbonic-acid gas.* The oxygen is the nourishment for the tissues,—the fuel for the fire; and the carbonic-acid gas is like the ash, which must be removed, or else the fire will be clogged. As the blood thus changes its load, a marked change in its color takes place. As it comes from the lungs, it is scarlet, and so continues through the heart and arteries until it reaches the capillaries: there it turns blue, and so continues through the veins and the heart and the pulmonary artery until it reaches the lungs again.

In the capillaries of the *larger, or systemic, circulation,* the blood turns *blue.*

In the capillaries of the *lesser, or pulmonary, circulation,* the blood turns *scarlet.*

3. We know that it is the oxygen taken in by the lungs that gives blood its scarlet color, and that it is the loss of oxygen in the tissues that makes it turn blue. For if we take some blue blood from a vein, and shake it up in the air, it will turn scarlet; and the color will be the same

when it contains much carbonic-acid gas as when it contains little.

4. The blood is the common carrier of the nutriment that we get from food and air; and it takes away, not only carbonic-acid gas, but all other waste matters. Some it discharges by the kidneys, some by the skin; the carbonic-acid gas, chiefly by the lungs.

WASTE MATTERS GIVEN OFF BY THE LUNGS.—CHANGES IN THE AIR.—VENTILATION.

SECTION VI.—1. The lungs, therefore, serve not only to take in oxygen, but also to discharge carbonic-acid gas, which is one of the chief waste products of the body.

Another product of the waste of the body, which is discharged by the lungs, is water. This ordinarily passes off in the breath as invisible vapor. But in very cold weather it is condensed as it comes out, and we can see it. It lodges and freezes on the beards of men, and on the hair of animals. If the vapor in the breath of a man is collected for twenty-four hours, and condensed to water, it measures as much as a pint.

2. The breath of man, and of every animal, when breathed out, contains, besides carbonic-acid gas and water, a very little of a certain substance which gives to each its peculiar odor. The breath of a cow, for example, has a smell peculiar to the animal. In pure, fresh human breath, we do not recognize any odor; but the substance is present: and in ill-ventilated rooms, where many people have been breathing, it becomes changed, and makes the air offensive and unhealthy. These three substances — oxygen, carbonic-acid gas, and an unnamed animal substance — are always present in

the breath. Other matters which give a distinct smell are often found in it, but not constantly.

3. The amount of air in a single breath is about twenty cubic inches, which would be a globe of air a little smaller than a base-ball. We take about eighteen breaths in a minute; in an hour, we breathe about twelve cubic feet of air; and in a day, about three hundred. Since we breathe more than eighteen times in a minute when we are exercising actively, probably three hundred and fifty feet would be nearer the amount used in a day. This would equal the entire contents of a room between seven and eight feet on every side.

4. Every breath, as it comes from the mouth, is changed in four ways,—

1. It has *lost* oxygen.

2. It has *gained* carbonic-acid gas.

3. It has *gained* watery vapor.

4. It has *gained* a nameless animal substance.

This breath is not fit to be breathed again. To be sure, there is still oxygen in it, but not so much as in fresh air. The loss of even one of the twenty-one parts in a hundred of oxygen, which fresh air contains, makes air less supporting; and if eleven parts are lost, though there are still ten parts left, we could not live in it. Moreover, the carbonic-acid gas has an injurious effect.

5. The stream of air which we pour out at each breath does not remain distinct, but mixes immediately with the surrounding air, just as a glass of colored fluid, if poured into a pail of water, will quickly diffuse itself through the whole. We never breathe precisely the same breath a second time, and the air of a room only very gradually becomes bad when many people are breathing in it. In

a large room, such as a church, many people can remain for a long time without suffering, even if the air is not changed. They may not be aware that it is growing impure, though one coming in from out of doors would know it at once. If the room is small, the air must be constantly changed, or they will be distressed.

6. The Black Hole, at Calcutta, was a room in which one hundred and forty-six men were confined over night. It was eighteen feet square, and had two small windows, through which the air did not come freely. Their sufferings were intense, and only twenty-three of them were alive in the morning.

7. Nothing is so free and abundant as fresh air; and yet we all suffer frequently, and many suffer constantly, for want of it. We should always remember its importance. It is possible to become accustomed to close rooms. We may not know what is the cause of a drowsy head, a disinclination to exercise, and general ill health, while it is simply lack of fresh air. Sleeping-rooms, in which we spend a third of every day, unconscious of the state of the air, should be well ventilated.

8. It is very difficult to get fresh air enough into our dwellings and public buildings, and, at the same time, to avoid draughts, and keep warm. *Ventilation* is almost a science by itself. There are many ways, but no one way which will do for every room. But if we all bear in mind the necessity of fresh air, and use our judgment and ingenuity, we shall get it.

THE VOICE.

SECTION VII. — 1. The organs of breathing are, at the same time, organs of voice. The *larynx* is especially the voice-box; but the lungs, the bronchial tubes, the trachea, and the throat, mouth, and nose, all have a part in making and forming sounds.

2. The toy pipes, which have, near the mouth-piece, narrow bands of brass, between which the air passes, and which make a reedy sound, are somewhat like the organ of voice. The chest and lungs give the wind, the trachea is the pipe, the larynx contains the vocal cords, which make the sound. The larynx is divided in two parts by a membranous partition, which has a slit in it, running forward and backward. This slit may be opened or shut, or made longer or shorter, by the action of many little muscles, which surround it.

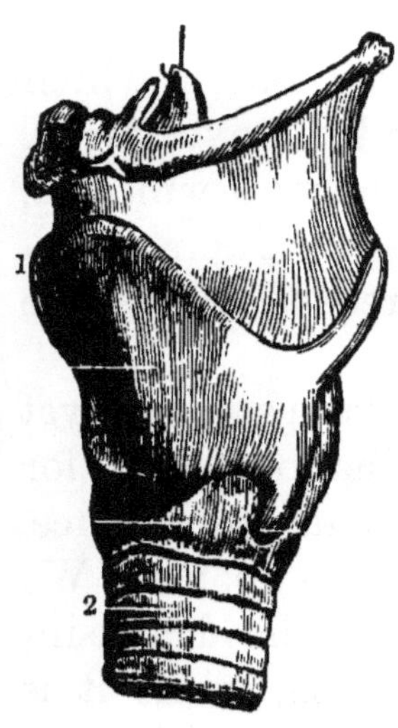

Fig. 52.

THE LARYNX.—1. Adam's Apple. 2. Trachea.

The edges of this slit, which is called the glottis, are the vocal cords. Every breath has to pass through it; but, in ordinary breathing, no sound is produced. If we wish to make a sound, the little muscles tighten up the cords, and make the slit narrow; and, as the lungs are squeezed in the chest, the air forced out through the slit makes the cords vibrate. This makes the sound. We shape this sound by changes in our throats and mouths.

3. A reed-organ is somewhat like the vocal apparatus; but, in the instrument, there is a pipe for every note. The windpipe can be so varied in length and in size, and the voice-box, with its cords, can be so changed in many ways, that a single pipe can make many different sounds.

4. The voice can be improved by cultivation. The breathing-muscles grow stronger, and also the muscles of the voice-box itself. The power to make firm, clear, sweet sounds increases. It is by singing that this is sought. But singing is not the only way of improving the vocal powers. To speak well is of more consequence than to sing well. In singing, the sound is the main thing, and less attention is paid to articulation. In speaking, it is important, not only that the tones should be clear and pleasing, but also that the pronunciation of words and syllables and letters should be distinct and correct. By taking care in speaking, we gradually train the muscles of the tongue and cheeks and throat, by which we make words, until it becomes a habit to speak well.

QUESTIONS.

SECTION I.—1. What is respiration?

2. What would air look like if we could see it?

3. Is air a material substance? What is it composed of?

4. What is the use of the nitrogen in the air?

5. What is the use of carbonic-acid gas in the air? Does man use it? What is adding constantly to the carbonic-acid gas in the air? What is adding constantly to the oxygen in the air?

6. Name some characteristics of oxygen.

SECTION II.—1. Does all of the air which we breathe in enter the blood?

2. What are the lungs? How does the oxygen pass through them into the blood-vessels?

3. How do frogs breathe under water?

4. How do fishes breathe?

5. In what are all breathing organs alike?

6. Illustrate the structure of a human lung by a bunch of grapes.

7. Illustrate the structure of a lung by a tree.

8. Can this structure be seen with the naked eye?

SECTION III.—1. Name the passages by which the air reaches the lungs.

2. What two parts does the nose consist of? With what other cavities are the cavities of the nose connected?

3, 4. Where are the nerves of smell, and how do the odorous particles reach them?

5. Give three reasons for breathing through the nose rather than the mouth.

6. What is a safeguard against taking cold when going from a heated room into cold air?

7. Where do the air-passage and the food-passage cross each other? Why does not food enter the windpipe?

8. What is the larynx?

9. Describe the trachea.

10. Into what does the trachea divide?

11. Describe the course and termination of the bronchial tubes.

SECTION IV.—1. Does air enter the lungs of itself?

2. What movements can we see when we watch a person breathing? What change takes place in the chest? Why does the air enter it? What is the chest like?

3. In what two ways is the chest made larger? Describe the first-named way.

4. Describe the second-named way.

5. Why do the lungs enlarge when the chest does?

6. What are the pleuræ?

7. Describe briefly the process of breathing.

8. Does the will have any thing to do with ordinary breathing?

SECTION V.—1. Follow the course of oxygen as it enters the blood.

2. What change in the color of the blood takes place? Where does the change from red to blue take place? Where does the change from blue to red take place?

3. What is the cause of the change in color?

4. What is the work of the blood?

SECTION VI.—1, 2. What do the lungs do besides taking in oxygen? What waste products are discharged by the lungs?

3. How much air is there in a breath? How much air do we breathe in a day?

4. What four changes take place in every breath? Has it lost all its oxygen? How much oxygen must air contain to support life? How small a loss of oxygen will be felt?

5. Do we breathe the same air a second time?

7. What are the immediate effects of lack of fresh air?

8. How should we try to secure fresh air?

SECTION VII.—1. What are the organs of voice?

2. How are the organs of voice like a reed-pipe? and what part does each do? What is the glottis? What are the vocal cords?

3. What resemblance between the vocal organs and a reed-organ?

4. How should the voice be cultivated?

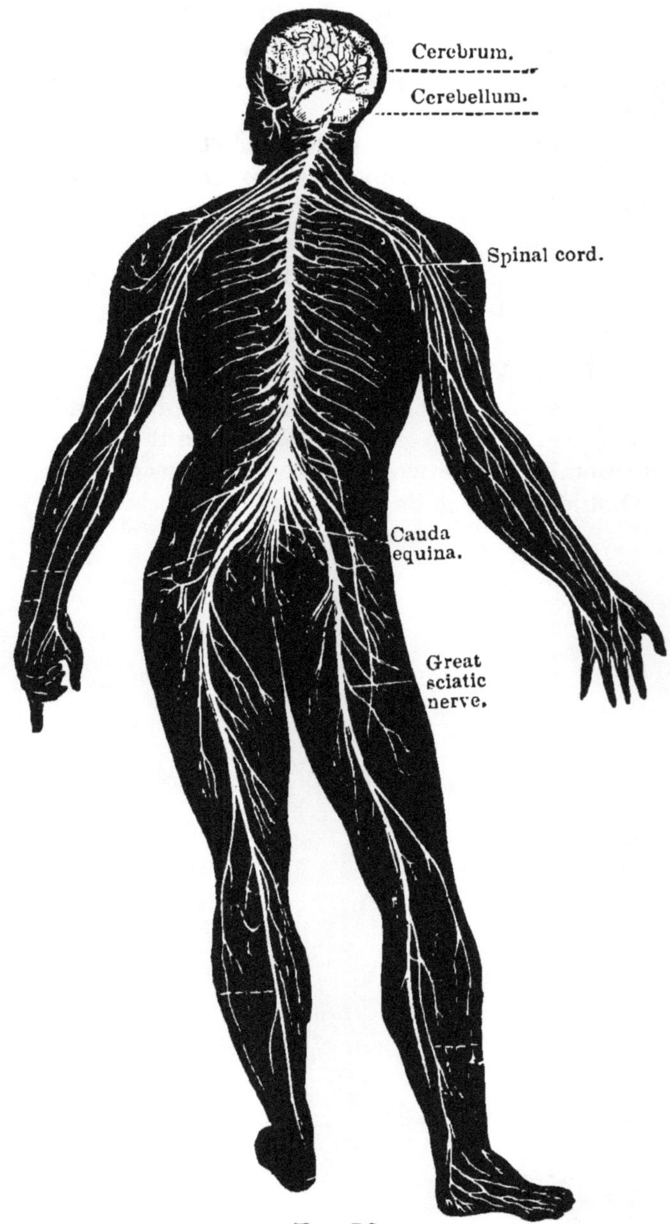

Fig. 53.

GENERAL REPRESENTATION OF THE NERVOUS SYSTEM.

CHAPTER IX.

THE NERVOUS SYSTEM.

SECTION I.—1. An injury to the head, if violent enough, will kill at once. If less violent, it will stun the victim. He will drop, limp and helpless, and will know nothing for a time; but his heart will continue to beat feebly, and he will still breathe. In time he will get his senses, and his power over his muscles again.

2. An injury to the backbone, if violent enough, will paralyze the lower limbs. The man is not stunned; and he breathes, and his heart beats. He can move with all the muscles above the injury: but those below are useless, though they have not been hurt; and perhaps he will have no feeling in those parts. In the case of the injury on the head, the part that is hurt is the *brain.* In the case of the injury on the back, the part that is hurt is the *spinal cord.*

3. The **brain** fills the chief cavity of the skull. It is not one mass, but several masses joined together. The largest mass is called the *cerebrum.* The next in size is the *cerebellum* (little brain), which lies behind and beneath the cerebrum. The *pons Varolii* is a mass in front of the cerebellum and beneath the cerebrum. The *medulla oblongata* is beneath the cerebellum, and behind the pons Varolii.

SUGGESTION TO TEACHERS.—SECTION I. 3. Get a sheep's or calf's brain. The chief divisions, the convolutions, and the gray and white matter, can be shown.

4. Each of these masses is in two halves, which are precisely alike. The two halves of the cerebrum and cerebellum are partly separate. The two halves of the pons and the medulla are united.

5. The surface of the cerebrum and cerebellum, instead of being smooth, is divided into ridges, with furrows between them. These ridges are called the *convolutions* of

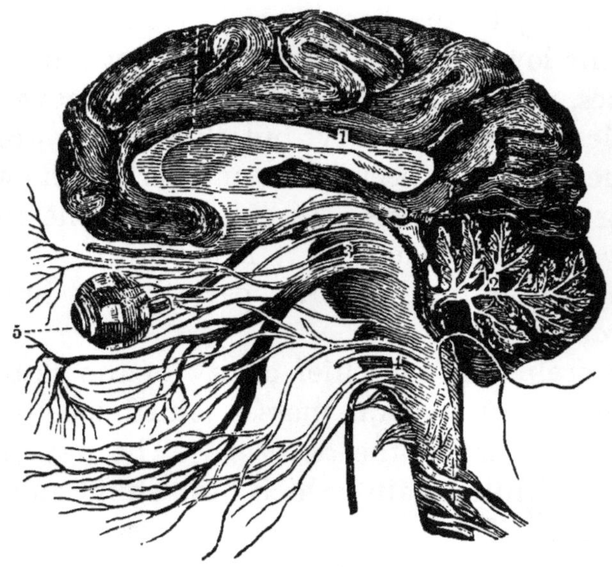

Fig. 54.

HALF OF THE BRAIN AND UPPER END OF THE SPINAL CORD, WITH THE NERVES COMING FROM THEM.—1. Cerebrum. 2. Cerebellum. 3. Pons Varolii. 4. Medulla oblongata. 5. The Eyeball.

the brain. One great difference between the brain of man and of the lower animals is, that it has more of these convolutions. Very intelligent animals have more of them than animals less intelligent.

6. The medulla oblongata is the lowest portion of the brain. It is just above the great opening in the base of the skull. Through that opening, it is continuous with the *spinal cord*.

7. The **spinal cord** occupies the spinal canal in the backbone. It is about half an inch thick, and eighteen inches long. Before reaching the lower end of the canal, it divides into a bunch of fine cords, which make the *cauda equina* (horse's tail).

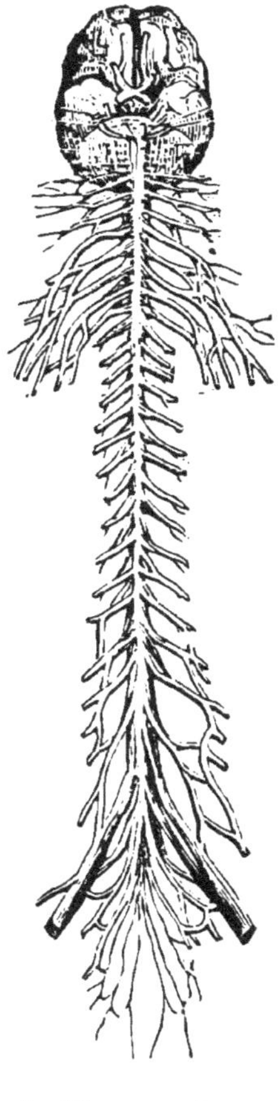

Fig. 55.

BRAIN AND SPINAL CORD SEEN FROM THE FRONT, WITH NERVES COMING FROM THEM.

The spinal cord, like the brain, is partly divided, lengthwise, into two halves, which are alike.

8. Both the brain and spinal cord have three coverings wrapped around them, called the *membranes* of the brain and cord.

From each half of the brain twelve small cords come off, and go out through holes in the skull. From each half of the spinal cord thirty-three small cords come off, by two roots each, and go out of the spinal canal through openings between the vertebræ. These are the **nerves**.

9. The substance of the brain and cord is soft and cheese-like. It is of two colors, white and gray. The whole surface of the cerebrum and cerebellum, about a quarter of an inch deep, is gray. The deep parts of both are chiefly white. In the cord, it is different. The deep part is gray, and the surface is white.

10. A blow on that part of the elbow called the "funny-bone," gives a tingling sensation all the way down to the little finger.

If we should cut into the elbow to find the "funny-bone," we would come upon a flat, shining cord about an eighth of an inch wide. If we should follow this cord down, we would find it divide into smaller cords; and, if we should follow each one of these, we would see it finally ending in the muscles, or in the skin, or some other tissue, in a great many fibers, so fine that only a microscope would show them. Many of them go to the little finger, where the tingling is most felt. If we should search farther for such cords, we would find that they were in nearly every part. We would find that some of them were as large as the one first seen, but that the smallest were smaller than the capillary blood-vessels, and as numerous.

11. If, starting from the elbow, we should follow the cord first found up the arm, we would see it joining other cords, and perhaps itself dividing; and finally we would trace it through one or more of the openings between the vertebræ (intervertebral foramina) into the spinal cord.

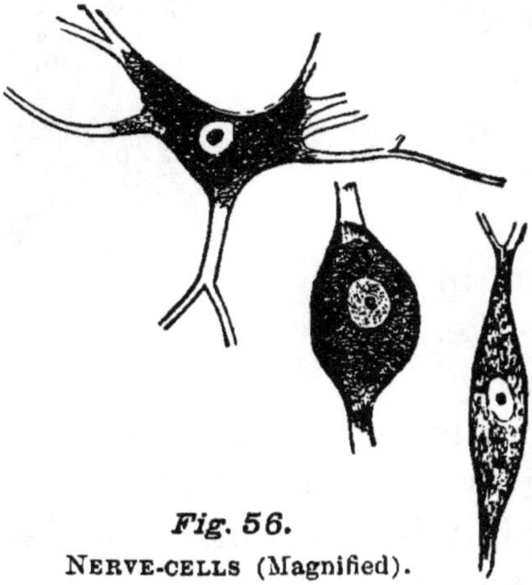

If we followed any other cord up, it would lead us finally into the spinal cord or the brain. Most of these white cords, which are the nerves, have one end divided into the finest of fibers in the different parts of the body, and the other end in the spinal cord or brain.

Fig. 56.
NERVE-CELLS (Magnified).

The brain and spinal cord are called the **nerve centers.**

12. By examination with the microscope, it is found

that the gray part of the brain and spinal cord is partly made of *cells* of various shapes, with fibers running out of them. The white matter is made of *nerve fibers* lying side by side. The nerves consist entirely of white nerve fibers. Most of the gray matter with its cells is found in the nerve centers.

ACTION OF THE NERVOUS SYSTEM.

SECTION II.—1. If nerves are cut off, we shall find a great change in the parts to which they go. If all the nerves going to a particular part of the skin are cut off, that part of the skin will be insensible. You may prick or pinch it, and it will not feel. If all the nerves going to a muscle are cut off, that muscle will not obey your will: you can no longer use it.

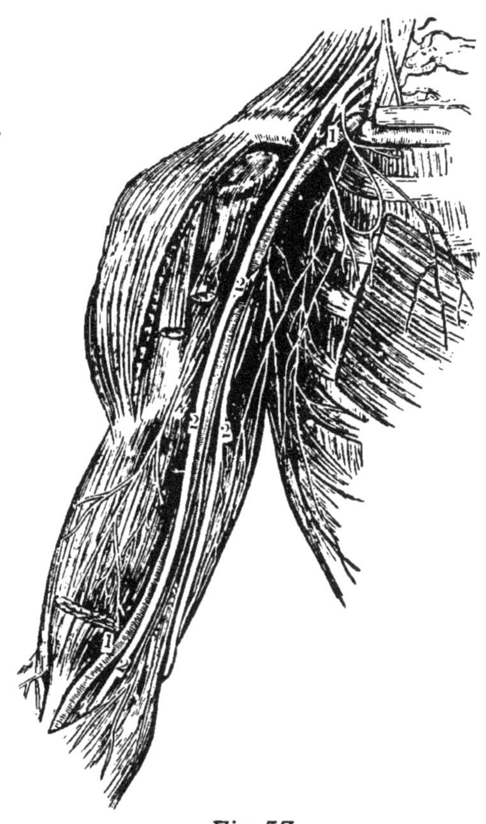

Fig. 57.

NERVES OF THE SHOULDER AND ARM.—1. Artery. 2. Nerves.

2. What you have done has not injured directly the skin or the muscle. It has cut off their connection with the nerve centers. We infer, therefore, that it is not really the skin that feels when we prick it. It is something in the nerve centers, and the nerve is the road by which the effect of the prick gets to the centers. When that road is cut off, we feel nothing. We infer, also, that the muscle does not commonly move of itself. It is caused to move by something in the nerve centers, and the nerve is the road by which the influence from the center that makes it move gets to it.

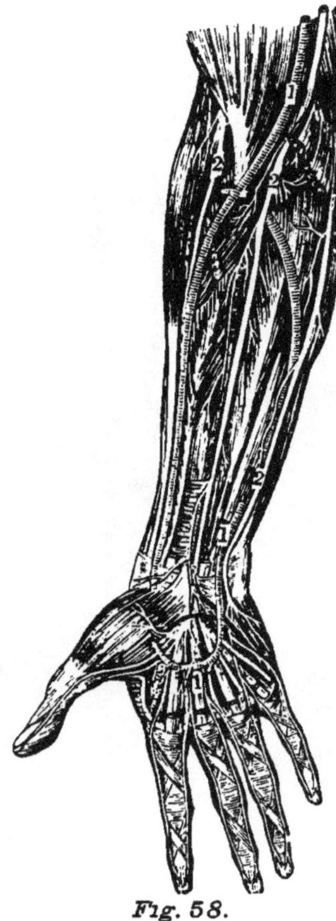

Fig. 58.

NERVES OF THE FORE-ARM AND HAND.—1. Artery. 2. Nerves.

3. The nervous system may be compared to the telegraphic system of a railroad. The nerves are the wires. The gray matter of the spinal cord contains the offices of the district superintendents. In the *cerebrum* is the office of the superintendent of the road. Suppose a mosquito lights on your face, and puts in his bill. He can not put it in, small as it is, without hitting one or more little nerve fibers. Instantly a message goes along those nerve fibers, through the nerve-trunks, through the spinal cord, and finally to the general superintendent's office. The message is, "Something wrong here." Immediately an order is sent out, along other nerve trunks and fibers, to the muscles of the shoulder and arm; and they contract, so as to strike the mosquito.

4. This is done very quickly, and yet it takes some time. The message goes in at the rate of one hundred and forty

feet in a second; it goes out at the rate of about ninety feet in a second; and it takes a little time for the superintendent to receive the message, and give the order. Then it takes time for the muscles to contract.

If the mosquito is quick, he will escape you.

REFLEX ACTION.

5. In the night, the general superintendent's office is closed: the cerebrum is asleep. And yet, if the foot is pricked or tickled, it will be drawn up. A frog's brain may be taken out entirely, without killing him. If then his side is pricked, he will scratch it with his foot. This shows us that the general superintendent in the cerebrum does not do all the regulating of the body. There are district superintendents in the spinal cord, and in the medulla; and some matters are never sent up to the central office at all. Digestion and respiration and circulation are all regulated by district superintendents.

6. Such actions are called *reflex actions;* because the message is sent in, and the order for the act "reflected" back, without any action of the will.

Winking is ordinarily a reflex act. The eye gets a little dry. The message goes in along the nerve fibers, which end just beneath its surface; and the order comes out to a muscle which brings the lids together, and spreads moisture over the eye.

Coughing is a reflex act. Something tickles the throat, or the air-passages below; and we can not help coughing. The muscles of expiration combine to throw out the thing that tickles.

Sneezing is a reflex act, for the purpose of clearing the air-passages above the throat. All these acts are regulated by centers below the cerebrum.

7. Many actions which must, at first, be attended to by the cerebrum, may, after a time, be handed over to the offices lower down. For example, in learning to play on the piano, it is necessary, at first, to give the whole mind to every touch. After a while, difficult music can be played with little thought.

If we were not made so that we could act in this way, our whole time and attention would be taken up in doing the things which now seem the simplest. Our minds would be worn out in making the movements, and combinations of movements, necessary in eating and breathing and walking.

8. The nerve-fibers which carry messages in to the centers are called *sensory fibers.* The nerve fibers which carry messages out to the muscles are called *motor fibers.*

9. The messages do not all come from the outside, as in the case of the mosquito-bite. A wish or thought may start in the cerebrum, without any impression from outside, and cause an order to be sent out to the muscles.

10. The nervous system connects the different parts of the body, and makes them work together.

THE CEREBRUM THE SEAT OF THE HIGHER FACULTIES.

Section III.—1. We have taken it for granted, that the cerebrum is the seat of the higher faculties,—the memory, the reason, the will, the feelings. This is proved in two ways:—

1. If this portion of the brain be diseased or injured, these faculties are affected.

2. The more of the higher faculties an animal has, the larger is the cerebrum in proportion to the rest of the brain. Man has the largest cerebrum proportionally.

SECTION IV.—1. We say that a person is *nervous* when his nervous system is excited by trifling matters. *Sickness* makes people nervous. *Close rooms* and *bad air* make people nervous. *Lack of exercise* makes people nervous. *Indigestible food* makes people nervous. The nervous system will endure a great deal of wear if rightly treated. Lack of sleep, too much excitement, and anxiety, often break it down. Worry is much more wearing than work. Stimulants, which spur the nerves continually, finally prostrate them.

2. Actions at first done with care and thought, by being done frequently, are at length done without effort. They become partly reflex acts. That is the reason why a skilled workman can work so much longer without being tired than a stronger man could, who was not so skillful. In this way *good habits* help us,—good habits of position, of movement, of speech, of study. So bad habits make slaves of us. The habit of biting the nails is, in many persons, a reflex act, which it is difficult to refrain from, because it is done without thought.

3. The brain needs exercise just as truly as the muscles. Study and thought not only make the brain strong and clear: they help to keep it in good health.

EFFECTS OF ALCOHOL ON THE NERVOUS SYSTEM.

SECTION V.—1. Alcoholic drinks are used chiefly for their effects on the nerves and brain. They stimulate the sense of taste in the tongue and palate, and they warm the stomach. Without really increasing strength, they give a feeling of strength and confidence. In moderate doses, they excite the brain. The sensibilities become lively, ideas flow readily, wit seems brighter, and philosophy more profound. The nerves that control the muscles are but little affected at this stage, and yet they are disturbed. A very little wine may spoil the chances of a rifleman in a shooting-match, or of a player in a game of ball.

2. As the amount of alcohol is increased, the higher faculties of the mind grow dull, while the lower propensities are still further excited. That portion of the brain which presides over the muscles loses its control. They still act with force; but they have no guide, and they do not act together. A drunken man can strike a hard blow, but he can not hit straight. The muscles of his legs fail to combine their action. They are all at cross-purposes, each contracting and relaxing without the direction of a central power. The man's movements become as tangled as his thoughts.

3. Vanity and pugnacity are now aroused; recklessness displaces caution; finally all self-control is lost, and the lowest instincts rule. Ungoverned impulses lead to crime and violence. The final stage is the drunken sleep.

The course of events varies with different temperaments. Some are but little excited, but gradually become stupid; some never lose control of their limbs; some are good-natured; and some uniformly morose. All are for the time insane.

4. The after-effects of free indulgence in alcoholic drinks are, an aching head, a foul stomach, unsteady nerves, and depression of spirits.

Drinkers often reach a condition in which this depression is constant, except when they are under the influence of liquor.

The appetite in many cases grows by gratification. It becomes so strong that it is almost impossible to resist it:

conscience is powerless; ambition, pride, and self-respect are abandoned. The most sacred affections are trampled under foot to satisfy the thirst. This condition is disease, but it is a disease for which the victim is himself responsible.

5. One of the most marked effects of alcohol is an enlargement of the small blood-vessels. This is what makes a toper's face red. Not only in the face, but in other parts, this enlargement takes place. Alcohol does it by paralyzing the little nerves, which are the regulators of the size of the vessels. These nerves constitute a very delicate mechanism of Nature's contriving, and it is important to the health of the body that it should not be interfered with.

6. Alcohol taken into the stomach is rapidly diffused through the body by the blood. Various experiments have proved that it accumulates especially in the brain. In this delicate organ, it causes not only the temporary effects already described, but permanent changes which manifest themselves in various diseases. Among them are epilepsy, paralysis, and insanity.

SUMMARY OF THE ACTION OF ALCOHOL ON THE BRAIN AND NERVES.

7. *1.* It first excites, and then paralyzes.

2. The higher faculties are the first to be paralyzed, leaving the man under control of his lower passions.

3. It obscures the senses, and impairs the judgment.

4. It exhausts the whole nervous system, and leads to paralysis, epilepsy, and insanity.

QUESTIONS.

Section I.—1. What may be the effects of injury to the head?

2. What may be the effect of an injury to the back?

3. Of what parts does the brain consist?

5. What are the convolutions of the brain? How do these indicate the difference between man and the lower animals?

6, 7. Describe the spinal cord.

8. What are the membranes of the brain and cord? How many nerves come from the brain? How many from the spinal cord?

9. What is the appearance of the brain substance? What is the appearance of the substance of the spinal cord?

10. Why does a blow on the funny-bone cause a tingling in the little finger?

11. What is the course of the nerves outward? What is the course of the nerves inward? What are the nerve centers?

12. What does the microscope show of the structure of the gray matter? Of the white matter?

Section II.—1. What is the effect of cutting off the nerve-supply of a portion of the skin? Of a muscle?

2. What do we infer from this effect?

3. Compare the nervous system to a railroad telegraph. What takes place when a mosquito stings?

4. Does it take time for a sensation to go through the nerves, or for a motor impulse to return? Do they go as fast as electricity?

5, 6. What are reflex actions? Name and describe some such actions.

7. May acts at first done by the cerebrum become reflex acts?

8. What are the sensory fibers? What are the motor nerve fibers?

10. How does the nervous system connect the different parts of the body?

Section III.—1. How do we know that the cerebrum is the seat of the higher faculties?

Section IV.—1. What is it to be nervous? What makes people nervous? What breaks down the nervous system?

2. How do good habits help us?

3. Does the brain need exercise?

SECTION V. — 1. For what chiefly are alcoholic drinks used? Do they increase strength? What is their first effect in moderate amount?

2, 3. What is the effect if drinking continues? Are all affected in the same way by alcohol?

4. What are the after-effects of free drinking? Is the appetite easily controlled?

5. How does alcohol make a toper's face red?

6. For what part of the body has alcohol a special affinity? What diseases of the brain are among its effects?

7. Give a summary of the effect of alcohol on the brain and nerves.

CHAPTER X.

THE SKIN.

STRUCTURE OF THE SKIN.

SECTION I.—1. *The skin* is the covering of the body. It is soft and smooth, but strong. It fits perfectly; but it stretches and glides a little on the muscles beneath, and, therefore, does not hinder our movements, as close-fitting garments do. It is partly transparent, and shows the blue color of the veins, and the red of the arteries, beneath it.

It becomes quite thick in places where a thick covering is needed, as on the palms of a laboring-man, or the soles of a barefoot boy.

2. If we prick a *blister*, a little watery fluid comes out, and it flattens down. It does not hurt to prick it, and it draws no blood. So we find that the top layer of the skin has no nerves or blood-vessels. But, when we "scrape the skin off," we are hurt, and have a red, bleeding surface. We do not often really scrape the skin off with trifling accidents; but we take off the top layer, and get down to the deep layer, which contains both nerves and blood-vessels.

3. *Dandruff*, which comes from the scalp, consists of dry scales from the surface of the skin. From all of the rest of the skin, little scales are constantly coming off.

SUGGESTION TO TEACHERS.—Illustrate this chapter by studies of the cuticle and hairs of men and animals, with a microscope. A lens of one-inch focus will show the pores on the palm.

They are much smaller than the scales of dandruff, and look like dust. We do not notice them, but they gather in our clothing; and we may scrape off some of them, and examine them with a microscope. They appear as dry white scales or flakes, and they are really dead skin. As this dead surface wears off, the deep layer must keep on growing, to make up.

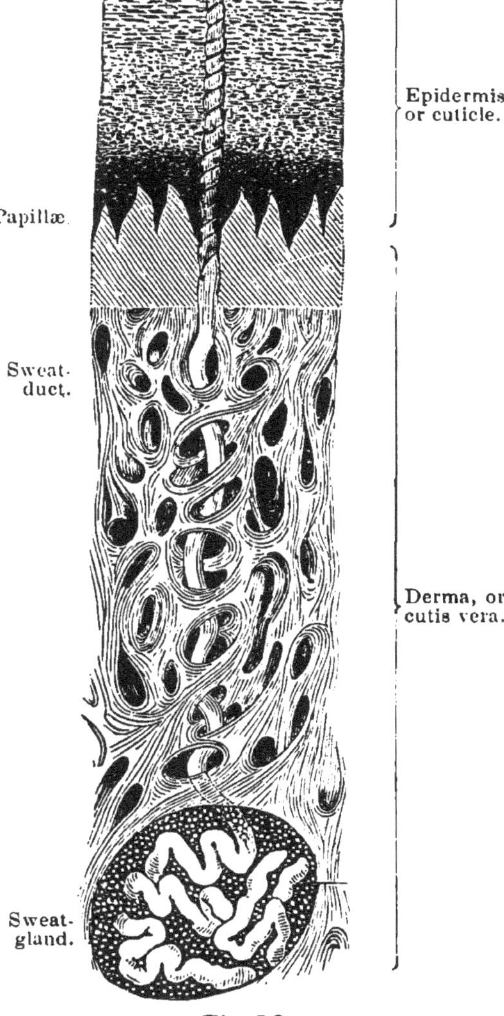

Fig. 59.
SECTION OF SKIN.

4. The top layer of the skin is called the *epidermis*, or *cuticle.* The deep layer of the skin is called the *derma*, or *cutis vera* (true skin). If a very thin slice of skin be taken out by cutting straight down through it, and we look at it with a microscope, we see something like the figure (Fig. 59). On the line which joins the derma and epidermis, there are little cones pointing up.

These are called *papillæ*, and are shown in larger size in Fig. 60, in which the epidermis has been stripped off, leaving only the derma with the papillæ on its surface. In this way we sometimes scrape off the epidermis by an accident, and these papillæ show as little red points. They contain blood-vessels and nerves.

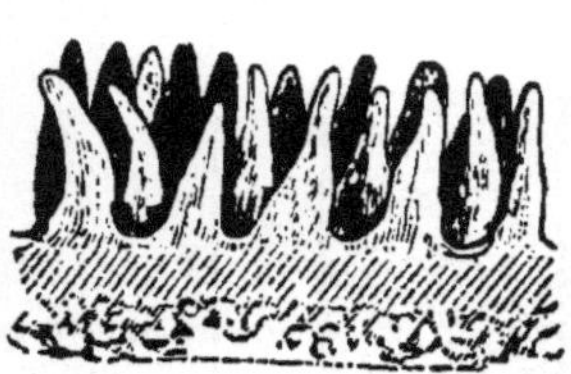

Fig. 60.
PAPILLÆ.

5. The palm of the hand, and especially the ends of the fingers, have distinct ridges, with furrows between them, which you can easily see. The ridges are made by the papillæ, which are numerous in these parts. When magnified, they look like the figure (Fig. 61).

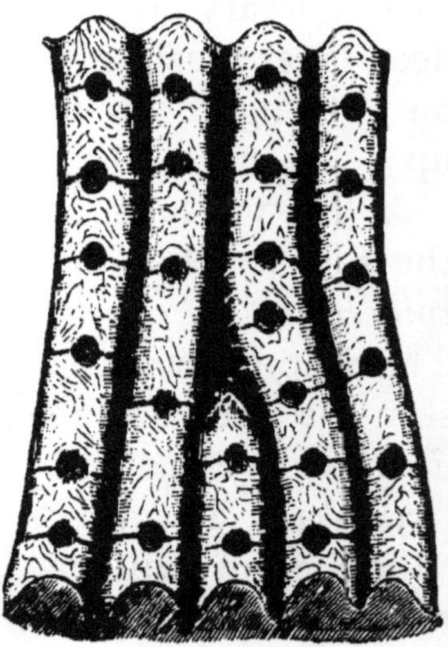

Fig. 61.
PAPILLARY RIDGES IN THE SKIN OF THE PALM. THE BLACK SPOTS ARE THE PORES.

6. The black spots on the ridges, in the figure, indicate the mouths of the sweat-ducts, which are called *pores*. In Fig. 59 we see one of these sweat-ducts. It is simply a tube. As we follow it down from the surface, it twists in corkscrew fashion through the cuticle, and then takes a wavy course through the cutis vera, and terminates in the deepest part of the skin, or just beneath it, in a coil, which is the *sweat-gland*. On the outside of this coil is a net-work of capillary vessels (Fig. 62). From these capillaries some of the water and salts of the blood pass through into the tube. As the tube fills, its contents well

up, and flow over on the surface. The tubes are like springs, drawing their supply from the blood-vessels beneath. It has been estimated that there are 2,500,000 of them altogether, and that, if they were all joined in one tube, it would be ten miles long.

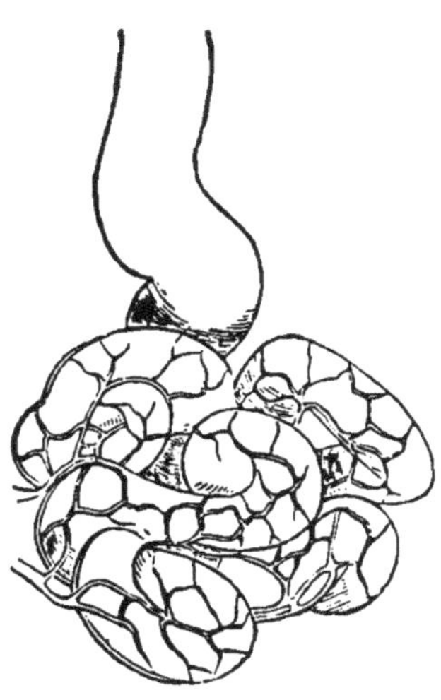

Fig. 62.
SWEAT-GLAND, WITH CAPILLARIES SURROUNDING IT.

PERSPIRATION.

7. We do not see any moisture on the surface when we are cool, but there always is some pouring out of the tubes. Ordinarily it changes into vapor, and is wafted away so fast that it can not gather in drops. But, when we are very warm, it wells up so rapidly that drops appear; and these sometimes flow in streams. If we remain quiet in a cool place, we dry off; that is, the perspiration does not come so fast, and that which is already on the surface is changed to vapor, and carried away by the air.

The moisture which is constantly coming from the pores, but which we can not see, is called *insensible perspiration.* That which we can see and feel, as water, is called *sensible perspiration,* or sweat.

HAIR.

8. Hairs grow from the skin. On most of the body they are short and fine. On the scalp and the face they grow long. The hair which covers the bodies of the lower animals is very useful in keeping them warm.

Man does not need it for this purpose, because he has intelligence to clothe himself. The head and throat are, however, protected in this way. In Fig. 63, we see the root of a hair as the microscope shows it. It is in the true skin, at the bottom of a tube. To the lower end of this tube, a muscular fiber is attached, which passes up to the surface of the skin. When this fiber contracts, it pulls up the hair, and makes the skin around it project, like a pimple. Cold makes these fibers contract: so does fear. It is in this way that the hair "stands on end," and that "goose-flesh" is made.

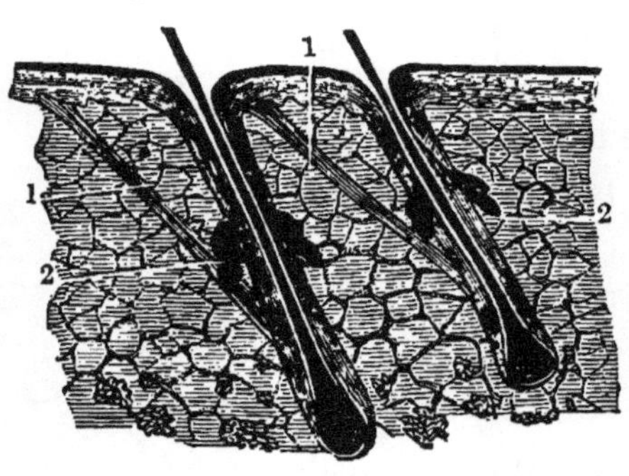

Fig. 63.

SECTION OF THE SKIN, SHOWING ROOTS OF HAIRS.—1. Muscles attached to the hair-sac. 2. Sebaceous glands.

SEBACEOUS GLANDS.

9. Opening into the tubes in which the hairs stand, are other tubes, which come from little sacs, called *sebaceous glands.* These lie in the skin, by the side of the hair-tubes. They manufacture an oily fluid, which is poured into the hair-tube, and out on the surface. It keeps the skin soft, and gives a moist and glossy appearance to the hair. When the scalp is unhealthy, and these glands are not active, the hair becomes harsh and dull and brittle.

NAILS.

10. The nails, like the hairs, grow from the skin. They are, indeed, a bit of the top layer, or cuticle, specially hardened. It is desirable that the ends of the fingers should

be firm for picking and touching; and, therefore, these little stiff backs are made to grow in them. The *root* of the nail is the upper end beneath the skin. The *matrix* is the bed on which it lies, and from which it grows. The nail may be shed or torn out, and, if the matrix is sound, a new nail will grow. If the whole matrix is destroyed, there will be no new one. But a very small portion of the matrix remaining, will produce a new nail, though it may be an imperfect one.

USES OF THE SKIN.

11. Understanding the structure of the skin, we are ready to notice its uses.

1. It *protects* the parts beneath. Being tough and elastic, it can bear hard knocks. The cuticle, which is without nerves, covers and guards the sensitive parts. Neither fluids nor gases pass through it easily. If the skin is whole, we can safely put our hands in poisonous fluids, which would enter the blood, and do us harm, if it were scratched.

2. It *gives off waste matters* from the body. Perspiration is chiefly water, but it contains some other substances dissolved in it. The amount of water given off in a day is different at different times and in different persons, but is ordinarily about a quart. In this amount of water, two or three spoonfuls of solid matter are dissolved.

3. It *regulates the heat* of the body.

BODILY HEAT.

12. Our bodies are always making heat; and yet, if we put the bulb of a thermometer in the mouth of a healthy man, it will never rise more than a degree or two above

98½° F. If it rises higher, he is sick, and has a fever. Nor will it fall more than a degree or two below 98½° F. In the stomach and in the blood, the thermometer would mark about 100° F., never rising much above, or falling much below. 100° F. is, then, the natural temperature of the inside of man's body. If he live under the tropical sun, it will not rise more than a degree or two above this if he is well; and, if he live in Greenland, it will not fall more than a degree or two below.

13. If we undertake to keep a room at a fixed temperature, we must have a fire to warm it; and we must have means of cooling it if it is too warm. The human body has its fire, and it has its cooling-apparatus; and this heating and cooling apparatus is self-regulating, and works so perfectly, that, throughout a long life, in heat and in cold, the inward temperature never varies more than two or three degrees in health.

HOW THE HEAT IS MADE EQUAL IN ALL PARTS.

14. The fire is not in any one part, but in all parts. In every particle, the changes which take place, as the particle takes up oxygen from the blood, and gives out carbonic acid, make heat, as the changes which take place in the coal or wood in the stove make heat. This heat is given to the blood. It is warmer when it comes from the capillaries, where it has been in close contact with the particles; and the blood, now divided in a thousand little streams, and again united in one stream, at the heart, diffuses the heat through the body.

15. Suppose the feet, for example, to be more exposed to cold than other parts. The blood in the feet might be, for a very short time, colder than the rest of the body;

but that blood immediately passes upward, and is warmed by the warmer stream above, with which it is mixed. A fresh supply of warm blood comes down, and contributes its heat to the feet. They are enabled in this way to maintain nearly the same heat as the rest of the body.

On the other hand, suppose the brain or the stomach, while they are especially busy, to grow hotter than the rest of the body. Their heated blood is soon mingled with the general stream; and, while it helps to heat the rest, is itself cooled. The heated parts are cooled by the cooler blood coming from other parts. So, by the constant circulation of the blood, the heat of all parts is made nearly equal.

16. The air is ordinarily cooler than the body, and is constantly taking heat from it. Clothing keeps us from losing heat too fast. Furs and woolens have no warmth in themselves. They only keep off the cold air, and keep in the heat that the body makes. Besides this, in cold weather, we warm the air by fires.

If we are much out of doors in cold weather, two things help us to keep warm: —

1. We *eat* more, and so furnish more fuel to our internal fires. We also get more oxygen in each breath, and this is fuel too.

2. We *exercise* more, and that keeps these internal fires more active.

In these ways, more heat is made; and we can afford to lose more.

HOW THE BODY IS COOLED.

17. Sometimes there is *too much heat* in our bodies, as when we are exercising, or in the sun. It will not go off into the air fast enough, even if we are lightly clothed.

In that case, the skin becomes a cooling-apparatus. It works in two ways: —

1. The deep layer of the skin is full of small blood-vessels. The effect of heat is to make these blood-vessels grow larger. The blood, then, flows into them, away from the deeper vessels. That is the reason your face gets red when you are heated. While the blood is in these vessels of the skin, it grows cool much faster than it does when it is deep in the body. It is nearer the cool air. The skin, then, receives more blood when we are heated, and spreads it out in a thin layer near the surface, and so cools it. By cooling the blood, the whole body is cooled.

2. The two or three million sweat-glands do a most important part of the regulation of the heat of the body.

In cities, water-carts go about in hot weather sprinkling the streets. This lays the dust, and cools the air. Much of the sprinkled water evaporates as it touches the warm stones; and, wherever water evaporates, it makes things around it a little cooler. The sweat-glands form a great watering-apparatus for the surface of the skin. The perspiration evaporates; and the skin, and the blood in it, are cooled. The hotter it is, the more we perspire; and, the more we perspire, the more heat is taken away. Men can stay for a time in a temperature of 200° F., and even more if they perspire freely. If perspiration is checked, they can not easily endure even a moderate heat. The reason why we suffer more from heat in what we call a "sticky day" in summer, is that the air is moist, and does not take up the moisture from the skin so fast as drier air would. Our bodies are wet, and evaporation goes on slowly.

Suggestion to Teachers. — Show the cooling effect of evaporation by throwing a spray of alcohol or ether on the hand, with an atomizer.

CARE OF THE SKIN.

Section II. — 1. If the skin of an animal is covered with varnish, it will soon die. If more than half of the surface of the skin is burned, even though the burn be not very deep, death will probably be the result. This shows how important the action of the skin is. It should be well taken care of.

2. To keep the skin healthy, three things are needed: —

1. To keep the *glands* (sweat-glands and sebaceous glands) *open and active.*

2. To keep the *blood circulating freely* in it.

3. To let the *air get to it.*

3. The solid matter in the perspiration, the oily matter from the sebaceous glands, and dead scales from the surface, together with dirt, will form a thin coating, which clogs the pores, and is itself unwholesome.

The skin of a savage is freely exposed to air. The civilized man, stepping from a warm bed into warm clothing, and staying, perhaps, much of his time in close rooms, gets a soft, over-sensitive skin. Neither its glands nor its blood-vessels are vigorous, and it can not do its work well.

4. Besides, such a man is liable to *colds.* A cold generally comes from chilling some portion of the skin. Those who are out of doors in all weather do not often catch cold. Their skin, as well as the rest of the body, can better resist a chill. Those who are seldom exposed to cold air, and whose skin has become delicate, catch cold most readily.

5. Too thick clothing keeps air from the skin, and weakens it by keeping it too warm. *Air-baths* are useful.

6. People who are sick or in pain, often receive great relief and comfort from having the skin rubbed; and it not only relieves, but, when thoroughly and perseveringly done, it helps to cure many diseases. It is good for well people also. *Thorough friction* of the whole body with a brush or a dry towel every morning, is the next best thing to a daily bath. It exercises the muscles of the skin. It brings the blood into it. It removes the accumulations of dead cuticle and perspiration.

7. No other one thing is so important for the health of the skin as bathing. It is possible for a delicate person to bathe too much, and imprudent bathing is sometimes hurtful. But, practiced with discretion, it is of great advantage. True, very many people enjoy good health who never bathe. But it seems unnecessary to argue, that to keep the skin clean, the pores open, the glands active, and the circulation free, by water and rubbing combined, must make the skin, and therefore the whole body, more healthy.

8. Cold water is a natural stimulant of the skin. The slight shock that it gives to the nervous system rouses the whole body to greater activity. Salt-water bathing is more invigorating than fresh-water bathing, because the salt has a direct, stimulating effect of its own.

9. There is no other safeguard against *colds* so good as a daily bath. We take cold because the skin is sensitive and delicate, and will not bear exposure to the damp or chilly air. By regular bathing, it is made vigorous, the blood flows freely through it, and its nerves are strengthened, so that we do not get chilled.

For that very common affection, *catarrh*, cold water, used freely and regularly on the skin, is an excellent remedy.

10. Certain cautions are to be observed in bathing: —

1. Never bathe *directly after* a full meal. The blood is then directed to the stomach, and active work is going on there. Other organs are relaxed. The bath will hinder digestion, and sometimes cause an injurious shock.

2. Never bathe in *cold* water when the system is greatly exhausted. Very delicate persons should not bathe in *cold* water at all.

3. Be very cautious about bathing in cold water *when heated*, especially if you are at the same time tired.

4. Never stay in the water until you are *chilled*, so that you do not get warm soon on coming out, or until you feel languid and weak instead of feeling refreshed.

Great injury sometimes results from neglect of these rules.

THE EAR.

SECTION III.—1. The ear, like the nose, is partly outside, and partly inside, of the head. It may be divided into—

The External Ear.
The Middle Ear.
The Internal Ear.

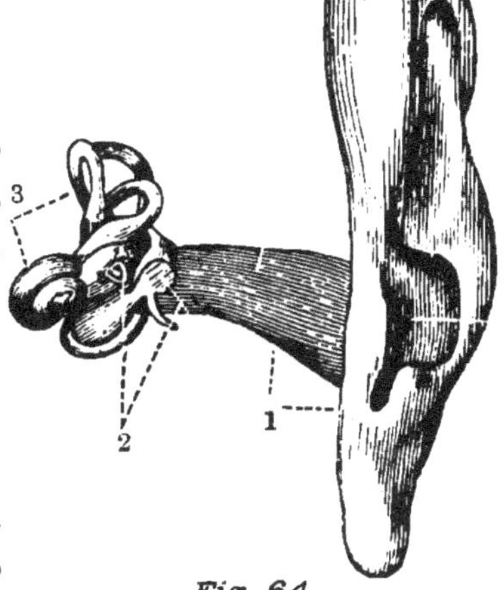

Fig. 64.
THE EAR.—1. Parts of the external ear. 2. Parts of the middle ear. 3. Parts of the internal ear.

2. The *external ear* is like an ear-trumpet; and its design is to collect the sound, and carry it in toward the internal ear, where the nerve of hearing is. Like an ear-trumpet, it has an open part and a tube. The tube enters the head: the open part is on the outside. It is

made chiefly of cartilage. There are three little muscles attached to the external ear, and some people can move it. Most people can not.

3. The tube is called the *external auditory canal.* It is an inch long, and ends at the membrane of the tympanum.

4. The *membrane of the tympanum*, or *drum-head*, separates the middle from the external ear. The cavity of the middle ear is called the *tympanum*, or *drum*. It contains the small bones of the ear. It is connected with the back part of the throat by a tube, called the *Eustachian tube.*

5. The *internal ear* is beyond the middle ear, deep in a bone of the skull. In it are the endings of the nerve of hearing.

HOW WE HEAR.

6. Sound is a vibration which can be perceived by the ear. Commonly as it reaches the ear it is a vibration of the air. The waves of air enter the external auditory canal, and strike the drum-head. They make the drum-head vibrate.

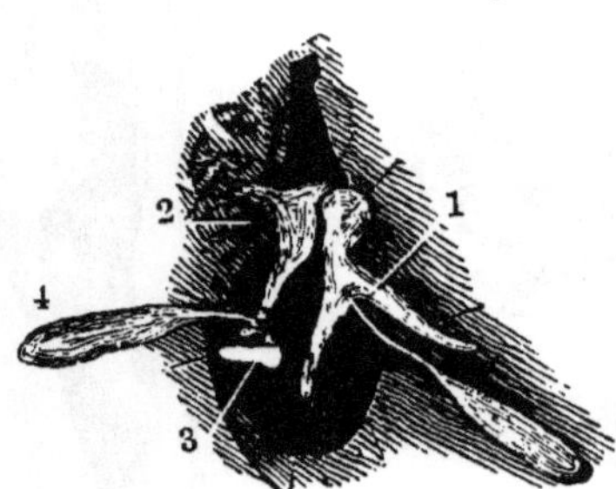

Fig. 65.

BONES OF THE LEFT EAR, SEEN FROM THE INSIDE. — 1. Hammer 2. Anvil. 3. Stirrup. 4. Stapedius.

7. Across the cavity of the drum, from the drum-head to the opposite wall, the three little bones of the ear — the "hammer," the "anvil," and the "stirrup" — are stretched in a chain. The hammer is joined to the drum-head and to the anvil, and the anvil to the stirrup.

8. When the drum-membrane vibrates, these little bones

are made to vibrate. The last one in the chain, the stirrup, is joined to a small membrane in the inner wall of the drum, which is like a little drum-head. On the other side of this little drum-head is the inner ear, which is filled with water. As the stirrup vibrates, it sets the little drum-head vibrating; and that makes the water in the inner ear vibrate, and the little waves strike the ends of the nerve of hearing, and by it the impression is carried in to the brain.

9. It is not the ear that hears. It is the brain that hears by means of the ear.

7. An *ear-ache* is commonly caused by inflammation of the lining of the drum. It swells, and discharges a fluid that fills the cavity, and makes pain by pressure. Sometimes, as the inflammation subsides, the fluid is absorbed. Sometimes the drum-membrane bursts, and lets out the fluid; and the pain stops. A discharge from the ear commonly comes from an inflamed middle ear through a hole in the drum-head. If the hole is small, it may heal up when the discharge stops. If a large part of the drum-membrane is gone, it will not heal up.

8. The loss of the drum-head does not destroy the hearing, but it impairs it.

9. *Ear-wax* is made by glands in the skin lining the auditory canal. It is not a safe practice to dig it out with hair-pins or other instruments. Ear-wax is necessary to keep the canal and drum-head soft and moist, and it will take care of itself. If it forms hard lumps, and stops the ear, as it sometimes does, it may be removed by syringing with warm water. No one but a physician should put in any instrument.

SECTION IV. — 1. The cavities in the skull which contain the eyes are called the **orbits**. They are shaped like pyramids, pointing inward, and are about an inch and a half deep. At the bottom of the orbits are holes, through which the nerves of the eye — the *optic nerves* — enter them. In the inner side of the orbits are openings into a canal, called the *lachrymal canal*, which runs straight down into the nose.

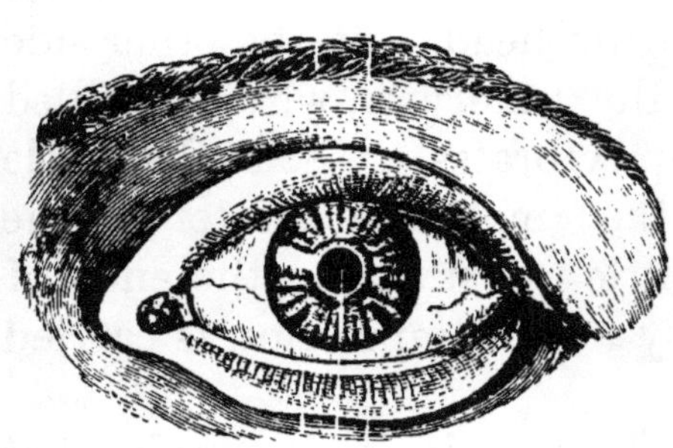

Fig. 66.
THE EYE.

2. The orbit is lined with fat, which makes a soft cushion for the eyeball. The **eyeball** is nearly round, and about an inch in diameter. It is attached to the optic nerve behind, as to a stem.

3. The outer coat of the eyeball is white and tough. It is called the *sclerotic* coat. A transparent circle, like a watch-glass, is set into this in front. It is called the **cornea.**

4. The cavity of the eyeball is divided into two chambers by the **lens**. The chamber behind the lens is filled with a jelly-like fluid, called the *vitreous humor*. The chamber before the lens is filled with a watery fluid, called the *aqueous humor*.

The optic nerve goes to the interior of the ball, and spreads out, by dividing up into fine threads, to line the back part. This lining is the **retina.**

SUGGESTION TO TEACHERS. — Get a beef's eye from the butcher's, and dissect it.

5. Looking through the transparent cornea, we see the **iris**, the colored part. It is very smooth and beautiful. It is made partly of muscle fibers, of the kind not subject to the will. Some of them form circles around its center: others run from center to edge, like the spokes of a wheel. When these latter fibers contract, the hole in the center, called the *pupil*, grows large. When the circular fibers contract, the pupil grows small. The iris is a curtain for the eye. Bright light makes it close up the pupil as much as possible. Opium and some other drugs will produce the same effect. Dim light makes it open the pupil. Belladonna produces the same effect.

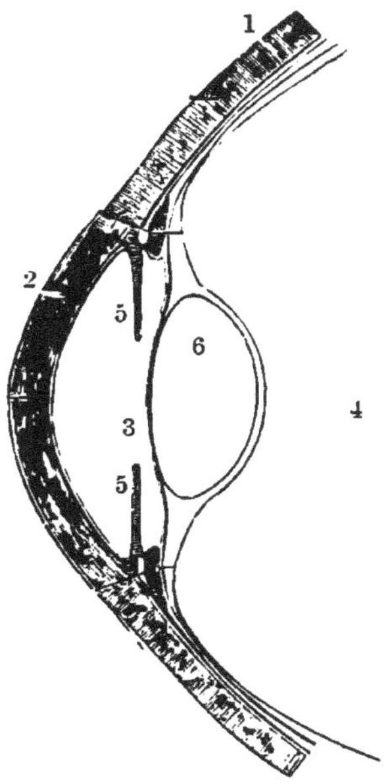

Fig. 67.

SECTION OF THE FRONT OF THE EYE.—1. Sclerotic. 2. Cornea. 3. Anterior chamber. 4. Posterior chamber. 5. Iris. 6. Lens.

The contractions of the iris are good illustrations of *reflex actions.*

6. The *lids* have each a thin plate of cartilage in them, to make them firm. The inside of each lid and the surface of the eye are covered with a mucous membrane, called the **conjunctiva.** The hairs (*eyelashes*) which grow from their edges help to protect the eye from dust and perspiration. Little sebaceous glands, called *Meibomian glands,* lie under the lining of the lids, and open on their edges. They oil the eyelashes and the edges of the lids. Under the roof of the outer part of the orbit, resting upon the eye-ball, is the tear-gland (*lachrymal gland*). About a dozen little ducts from it open

on the surface of the eye. This gland furnishes the moisture which the eye requires. By winking, the moisture is spread over the surface.

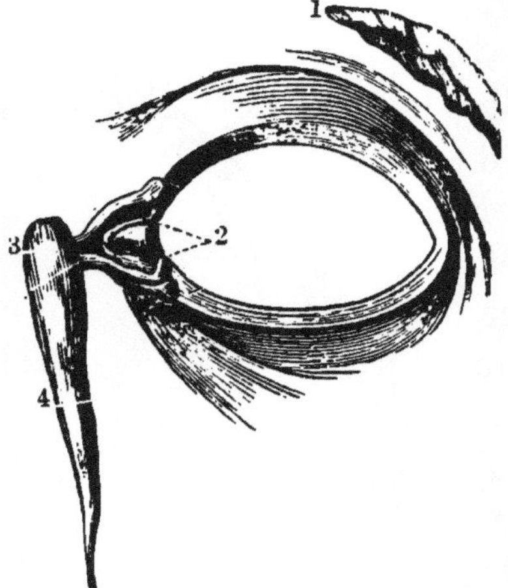

Fig. 68.

LACHRYMAL APPARATUS.—1. Lachrymal gland. 2. Tear-passages. 3. Lachrymal sac. 4. Nasal duct.

7. Ordinarily moisture is supplied only as fast as it is needed, and it all evaporates. In strong feelings of sorrow or joy, moisture is poured out very rapidly, and gathers in *tears*. Near the inner angle of the eye, an opening, apparently about as large as a needle, can be seen on the edge of each lid. These are the openings of the ducts which carry off the tears. They lead into a sac in the inner corner of the orbit (the lachrymal sac); and that opens into the *nasal duct*, which runs through the lachrymal canal into the nose. When tears are very abundant. they overflow on the face.

8. There are six *muscles* attached to the eyeball. One rolls it up, one down, one out, one in, and two roll it on an axis passing from before backward.

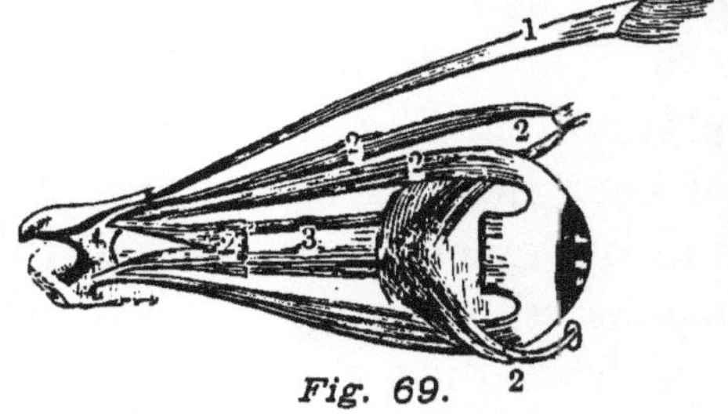

Fig. 69.

MUSCLES OF THE EYE-BALL.—1. Muscle of the upper lid. 2. Muscles of the ball. 3. Optic nerve.

HOW WE SEE.

9. The eye is like the camera with which a photographer takes pictures. The

SUGGESTION TO TEACHERS.—Show and describe the parts of a camera.

lens is like the glass lens in the end of the tube. The lining of the cavity of the eye-ball is colored dark, like the inside of the box of the camera. The retina is like the sensitive plate which the photographer puts in just before he takes the picture. The brain behind the eye is represented by the photographer himself looking through from behind his instrument.

10. When we look at an object, a picture of it is made on the retina. This picture can be seen by a skillful observer looking into the eye with an instrument called an ophthalmoscope. It may be seen by looking from behind through an eye taken from an animal just killed, as the photographer looks through his camera.

The retina, which is the ending of the optic nerve, carries in to the brain the impression which makes the picture. It is not the eye that sees. If the optic nerve is cut off, we do not see; although a picture will still be formed on the retina as before. It is the brain that sees. The eye is its instrument.

CAUSES OF TROUBLE IN THE EYE.

11. The eye is sometimes weak, and gets very tired and sore by use, simply because the whole body is weak. But generally when there is redness, swelling, or pain in the eye, there is something wrong in itself. It may be that the delicate lining of the lids, the *conjunctiva*, is inflamed, just as the lining of the throat is inflamed at times. The lids then feel rough, as if there were sand under them.

TOO LONG SIGHT.

12. One of the commonest causes of aching eyes and head, after reading, is *too long sight.* When we look at a near object, the shape of the lens is changed a little by an effort of the eye. Too long-sighted eyes have to make more effort to do this than eyes with natural sight. The strain tires and irritates them, and may make them very weak. This trouble can be entirely relieved by glasses.

TOO SHORT SIGHT.

13. Short-sighted persons do not commonly have aching and inflamed eyes. Their eyes do not have to make the effort that those of long-sighted people do to see near objects. Reading does not tire them. But short-sightedness is liable to increase. It makes the vision of every thing more than a few feet away indistinct. It may be remedied by glasses, but they are an inconvenient necessity.

14. Too short sight is very common among students. It is found, that, when children begin to go to school, few of them are short-sighted. In each higher class, there are more short-sighted pupils; and the number increases so fast, that we infer that there is something in the habits of school-children that makes them short-sighted.

CAUSES OF TOO SHORT SIGHT.

15. If we inquire what this cause of short sight is, we find that it is not any one thing, but many things. Every thing that tires and strains the eyes of school-children, tends to make them short-sighted.

Causes of short sight are,—

1. Too much use of the eyes.

2. Bad light.

3. Wrong positions when reading.

16. *1.* If the eyes are tired and hot, it is a sign that they have been used too long. *Stop* until they are rested.

2. Do not try to read when there is not light enough. If the light is dim, the book will be held too near the eyes. This tires them, and causes short sight. Reading by a *fading twilight* is particularly bad.

3. The light should not be *too glaring.* If it is, the nerve is stimulated too much. This tires it. Besides, when any nerve is over-stimulated, it loses its sensitiveness after a time; and then the book will be held too near, and short-sightedness may result.

For the same reason, the light should, if possible, fall over the left shoulder on the page. Then it will not be all reflected into the eyes, as when it comes from in front.

4. The *light* should be *steady.* A flickering light keeps the eye annoyed, and tires it with constant changes.

5. The *book* and the *eye* should be *steady.* It is as wearisome to the eye to have the page or the head in constant motion as to have the light flicker. Reading in the cars is trying to the eyes.

6. The *upright* position is the natural and easy one for the eyes. To read when lying down, or with the head hanging over the book, tries the eyes, and tends to short-sightedness.

7. Any serious trouble with the eyes should be *attended to at once.* It is better never to open a book than to lose the use of the eyes. If study can not be continued without ruining the eyes, abandon study.

COLOR-BLINDNESS.

17. Not a few people are color-blind. Some can not distinguish any colors. Others can not recognize a particular color, as red or blue, confounding it with other colors. If a person is color-blind, it is very desirable to know it. An engineer on a railroad, who could not tell a red light from a green one, would be a dangerous person.

QUESTIONS.

SECTION I.—1. Name some qualities of the skin.

2. What is a blister? Is the top layer of the skin sensitive? Are there any blood-vessels in the top layer? Are there any nerves or blood-vessels in the deep layer?

3. What is dandruff? Is there any thing similar to it from the surface of the whole body?

4. What is the top layer of the skin called? What is the deep layer of the skin called? What are the papillæ, and where are they situated?

5. What makes the ridges and furrows on the skin of the palm?

6. What are the pores? Describe a sweat-duct. Describe a sweat-gland. Where does the sweat come from? How does it get into the duct? What is the estimated number of pores? What is the estimated length of all the ducts combined?

7. What is insensible perspiration? What is sensible perspiration?

8. What is the use of the hair? What does it grow from? How do the muscular fibers of the skin act on the hairs?

9. Where are the sebaceous glands, and what do they do?

10. What is the use of the nails? What do they grow from? What is the root of the nail? What is the matrix of the nail? When a nail comes off, is it ever restored? When is it not restored?

11. Name three uses of the skin.

12. What is the natural heat of the inside of the body?

13. How is it kept just at this temperature?

14. What makes heat in the body?

15. How is the temperature of the different parts kept equal?

16. How do clothes keep the body warm? In what two ways are we helped to resist the cold?

17. Describe the first-named method in which the skin acts as a cooling-apparatus. Describe the second-named method in which the skin acts as a cooling-apparatus. How can men endure a temperature of 200° F.?

SECTION II.—1. What shows the importance of the action of the skin?

2. What three things are necessary in order to keep the skin healthy?

3. How does the skin become clogged and over-sensitive?

4. How may we become liable to colds?

5. What is the harm of too thick clothing?

6. What are the advantages of rubbing the skin?

7. What is the use of bathing?

8. Why is cold-water bathing invigorating?

9. What is the best safeguard against colds?

10. What cautions are to be observed in bathing?

SECTION III.—1. Name the three divisions of the ear.

2. What is the use of the external ear? What is it made of?

3. Describe the external auditory canal.

4. What is the drum-head? What is the drum? Is there any opening out of the drum?

5. Where is the internal ear? What is its importance?

6. What is sound?

7. What is the position of the little bones in the ear?

8. What is the use of the little bones? How does the vibration reach the nerve of hearing?

9. Does the ear hear?

10. What frequently causes an ear-ache? From what spot does a discharge from the ear commonly come? Is a broken drum-head ever repaired?

11. Does the loss of the drum-head destroy hearing?

12. What is ear-wax? Is it safe to put instruments in the ear?

SECTION IV. — 1. Describe the orbits.

2. State the shape, size, and position of the eyeball.

3. What is the sclerotic coat of the eye? What is the cornea?

4. What divides the eye into two chambers? What fills the posterior chamber? What fills the anterior chamber? What is the retina?

5. Describe the iris. The pupil. What is the action of the iris?

6. What is the conjunctiva? What is the use of the eyelashes? What are the Meibomian glands? and what is their use? Where is the lachrymal gland? What is its use?

7. What are tears? How are they carried off from the eye?

8. What muscles move the eyeball? and how do they act?

9. How does the eye resemble a photographer's camera?

10. Where is a picture of the object looked at formed? Does the eye see?

11. Name some causes of trouble in the eye.

12. What is the effect of too long sight? What is the remedy?

13. What is the effect of too short sight?

14. Among what class is it most common?

15. What are causes of too short sight?

16. Give rules for the care of the eyes.

17. What is color-blindness?

APPENDIX.

APPENDIX.

WHAT TO DO IN CASE OF ACCIDENT.

It is natural to be alarmed when an accident occurs. Our feelings of sympathy for the sufferer, and perhaps of fear that he will die, agitate us, and scatter our wits, and make us helpless. The first thing at such a time, is to *think*. "What can I do?" is the question. To apply the mind vigorously to that question, is the best way to control the feelings.

We can not anticipate all the particulars of accidents that may happen, but we can fix in the mind a few simple directions for each kind.

Fainting. — When a person faints, the heart almost stops beating. The face is deadly pale. If we could see the brain, we should see that pale too. Because the blood is not sent to it in sufficient amount, it partly stops acting, and the person is unconscious.

Place him on his back, with his *head low*. The blood will flow to the brain more easily in the horizontal than in the upright position.

Give him air. Perhaps he has fainted because the air is bad. Fresh air will revive him.

Sprinkle cold water on the face. It stimulates and rouses the nerves.

Loosen the clothing about the neck and waist, so that it may be easy to breathe.

A fainting-fit generally lasts only a short time.

Fits. — In fainting, the face is pale, the pulse can hardly be felt, the limbs are limp and still. In *fits*, the face may be pale or red, the pulse can be felt easily, and the limbs often jerk and draw up spasmodically. Frothing at the mouth is not uncommon. If the fit is a long one, or if several come in succession, there will be time for the doctor to arrive. What you can do, is, —

1. To *keep the person from hurting himself* in his struggles.
2. To give him *plenty of air*.
3. To see that there is *nothing tight* around his neck or chest.
4. Place him with his head *raised* a little.
5. When he comes out of the fit, let him *rest*.

Sunstroke. — This happens to those who have been exposed to great heat, either in the sunshine or in the shade. The face is flushed or pale, the pulse quick, and the skin dry and hot.

Put the sufferer in as cool a place as possible, with his head raised, and apply *cold water or ice* to his head and chest. If he seems extremely weak, and the skin becomes cool, stop using ice, and put mustard and water on his feet and on the back of his neck.

Shock. — Shock is the name given to the condition of prostration which sometimes follows a severe injury. The person is conscious, but extremely weak. The face is pale, the skin cool, and perhaps moist, the pulse quick and small, and there is restlessness. The powers may continue to fail until death comes. More frequently they rally after a time.

A person in this condition must be handled very carefully. Any roughness may quench the spark of life. He must be laid down with his head low. Nature must be aided by gentle stimulants. Heat is one of the best stimulants. Put hot-water bottles at his feet and sides. Give air, but avoid chilling him. Do not move him until he is better.

Fractures and Dislocations. — A broken or dislocated limb is generally helpless. Even when there is no pain in it, it can not be used. If there is reason for thinking that such an accident has happened, the limb should be seen immediately by a doctor. In the mean time keep it perfectly quiet. Do not allow the patient to attempt to use it. Support it in the position in which it is most comfortable. Bathe it in cold or hot water to relieve pain and keep down swelling.

Bleeding. — When blood is flowing from a wound, it must be stopped. In most cases it is only necessary to press a handkerchief on it. This will check the flow so that the blood can coagulate in the vessels. Sometimes it is necessary to press hard upon the wound for a good while. Sometimes this will not be sufficient. The blood is flowing, perhaps, from a large artery. Then a bandage must be drawn tight around the limb *above* the wound. If the bleeding stops, the bandage may be taken off in half an hour. If the bleeding does not stop, it is, perhaps, coming from a vein; and a bandage must be tied around the limb *below* the wound. It may be necessary to find the course of the bleeding vessel, and press on it with the thumb. In some of these ways, external bleeding can always be checked for the time.

Bleeding from the Nose. — Let the person keep upright, and hold his hands above his head. Apply cold to the back of the neck and to the forehead. Press on the nostrils. Push a little wad of cotton into the nostrils.

Bleeding from the Lungs or Stomach or Bowels. — Keep the patient perfectly quiet on his back. Do not let him talk. Give him ice to swallow, and salt and water.

Poisoning.—If the poison has not been in the stomach more than an hour or two, try to make the patient vomit. This can sometimes be done by tickling the throat with the finger or a feather. If this is not sufficient, give a teaspoonful of mustard in a tumbler of lukewarm water. Or give a dose of ipecac. If the poison is causing pain in the stomach, give the whites of two or three eggs. If the poison is an acid (as sulphuric acid), give soapsuds or magnesia. If the poison is a narcotic, like opium or belladonna or chloral, keep the person from going to sleep.

Drowning.—When a person, apparently drowned, is taken out of the water, his lungs have water in them, and his throat is stopped with water and mucus. Turn him on his face, and let the water run out of his mouth. Thrust your finger in his mouth, and clear it, and draw his tongue forward. Then lay him on his back, with a folded coat under his shoulders, raising them a little higher than his head. Tickle his nostrils with a feather. Give a little snuff, or a smell of smelling-salts. This may start his breathing. If it does not, try to make his chest expand and contract as it does naturally. A good way to do this is to stand across his body, or above his head, take hold of his arms, and bring his elbows up to the sides of his head. This opens the chest and lungs, and imitates the movement of inspiration. Then bring his arms down to his sides, and press them against the walls of the chest. This contracts the chest and lungs, and imitates the movement of expiration. Do this fifteen or twenty times in a minute for half an hour, if the chest does not begin to open and contract of itself in less time. At the same time, let others get off the wet clothing, and cover with dry. Let them rub the limbs vigorously, rubbing from the extremity toward the center. If bottles of hot water, or hot bricks or hot flannels, can be had, put them at his feet and sides. The three things to be aimed at, are,—

1. To start the movements of breathing.

2. To restore warmth.

3. To quicken the current of blood.

Persevering and energetic efforts will often be rewarded by a return of the signs of life when they seemed to have gone forever.

Remember that "a little knowledge is a dangerous thing" if it leads to undue self-confidence. In case of serious accident, never fail to get help if possible from some one more experienced and capable than yourself.

GLOSSARY.

GLOSSARY.

Ab-dō'men. The belly.

Ad'am's ap-ple. A prominence in the middle line of the neck.

A-dŭl'ter-āte. To make impure by adding inferior substances.

Al'i-mĕnt. Nourishment.

Al-i-mĕnt'a-ry. Pertaining to aliment.

A-năt'o-my. The science of the structure of organized bodies.

A-or'ta. The great artery which comes from the heart, and passes down by the backbone.

Ap-o-neŭ-rō'sis. A membrane of white fibrous tissue connected with a muscle.

Ap'o-plĕx-y. A disease of the brain in which sense, and power of motion, are suddenly lost.

A'que-oŭs. Watery. A term applied to the fluid contents of the anterior chamber of the eyeball.

Ath'lēte (Lat., *athleta*, one who contends for a prize). One who especially cultivates his muscles.

Au'di-to-ry (Lat., *auditus*, hearing). Pertaining to the hearing.

Au'ri-cle (Lat., *auricula*, a little ear). A name given to two of the cavities of the heart.

Ax-ĭl'la. The arm-pit.

Bī'ceps (Lat., *bis, caput*, two-headed) A muscle extending from the shoulder to the fore-arm, on the front of the arm. Also, a muscle extending from the hip to the leg, on the back of the thigh.

Brăch'i-al (Lat., *brachium*, the arm). Pertaining to the arm.

Bŭn'ion. An enlargement and inflammation at the first joint of the great toe.

Căp'il-la-ry (Lat., *capillus*, hair). Hairlike

Ca-rŏt'id. A name applied to several arteries in the neck.

Car'pus. The wrist

Car'ti-lage. Gristle

Ca-tarrh'. An inflammation of the mucous membrane.

Cau'da e-quī-na (horse's tail). The bundle of nerves into which the spinal cord divides at its lower end.

Cĕr-e-bĕl'lum. A division of the brain, behind and beneath the cerebrum.

Cĕr'e-brŭm. The largest division of the brain.

Chȳle. The emulsion of fats made in the intestine.

Chȳme. The gruel-like mixture which passes from the stomach into the intestine in digestion.

Clăv'i-cle. The collar-bone.

Co-ăg-ū-lā'tion. The forming of clots

Cŏc'cyx (Lat., *coccyx*, a cuckoo) A bone at the lower end of the backbone, shaped like a cuckoo's beak.

Con-jŭnc-tī'va. The mucous membrane covering the front of the eyeball, and lining the lids.

Cŏn-vo-lū'tions. Ridges on the surface of the cerebrum and cerebellum.

Corn (Lat., *cornu*, a horn). A small portion of the epidermis, of hornlike hardness.

Cor'ne-a (Lat., *cornea*, horny). The circular, transparent membrane in the front of the eye.

Cor'pus-cle (Lat., *corpusculum*, a small body). A minute particle

Cŏs'tal (Lat., *costa*, a rib). Pertaining to the ribs.

Cū'ti-cle. The upper layer of the skin.

Cū'tis ve'ra (true skin). The deep layer of the skin.

Dăn'druff. A scurf which forms on the scalp, and comes off in small scales.

Dĕn'tīne (Lat., *dens*, a tooth). A bonelike substance of which the teeth are made.

Der'ma. The deep layer of the skin.

Dī'a-phrăgm (midriff) A sheet made of muscle and fibrous membrane between the chest and abdomen.

Dŭct. A tube by which a fluid, or other substance, is conducted.

Dys-pĕp'si-a. Bad digestion.

El'e-ment. One of the simplest parts of which any thing consists. A name applied to those simple substances, between sixty and seventy in number, to some of which all material objects can be reduced.

E-mŭl'sion. A mixture of oil with water containing some gummy or albuminous substance.

En-ăm'el. The hard and polished substance which covers the crown of a tooth.

Ep-i-dĕrm'is. The upper layer of the skin.

Eū-stā'chi-an tube. A tube connecting the middle ear with the throat.

Ex-pī'ra-to-ry (Lat., *ex spirare*, to breathe out). Out-breathing.

Fĕm'o-ral. Pertaining to the thigh.

Fē'mur. The thigh-bone.

Fī'ber. A thread of tissue.

Fī-bro-car'ti-lage. A tissue made of cartilage, with white fibers mixed with it.

Fī'broŭs mĕm'brāne. A membrane made of fibers.

Fĭb'ū-la. A slender bone in the calf of the leg.

Flā'vor. That which gives a peculiar odor or taste.

Fŏs'sa (Lat., *fossa*, a ditch). A depression in a bone.

Găs'tric. Pertaining to the stomach.

Găs-troc-ne'mi-us. The large muscle extending from the thigh to the heel on the calf of the leg.

Glănd. A name applied to many organs which take part in the processes of life.

Glŏb'ūle. A little globe.

Glŏt'tis. A slit in the membranous partition between the upper and lower parts of the larynx.

Glū'ten. A substance in grain that contains the same chemical elements that meat contains.

Gly'co-gen. A substance formed in the body chiefly by the liver.

Ha-ver'sian ca-năls' (from Havers, who first described them). Microscopic canals in bone, in which the blood-vessels run.

Hū'me-rŭs. The arm-bone.

Hȳ'gi-ēne. The science of health.

Hȳ'oid. U-shaped.

In-spī'ra-to-ry (Lat. *In spirare*, to breathe in). In-breathing.

I'ris (Lat., *iris*, the rainbow). A colored muscular membrane in the anterior chamber of the eye.

Jaun'dice. A disease in which the body is colored yellow.

Jū'gu-lar (Lat., *jugulum*, the neck). Pertaining to the neck.

Lăch'ry-mal (Lat., *lacryma*, a tear). Pertaining to tears.

Lăc'te-al (Lat., *lac*, milk). A term applied to the lymphatic ducts of the intestine.

Lĭg'a-ment (Lat., *ligare*, to bind). A fibrous band that binds two parts (commonly bones) together.

Lȳmph. The contents of the lymphatic vessels.

Măr'rōw. A soft substance contained in the cavities of bone.

Mā'trix. An organ which produces or gives form to any thing.

Me-dŭl'la ob-lon-ga'ta. The lowest division of the brain.

Meĭ-bō'mi-an. Discovered or described by Meibomius.

Mĕt-a-car'pus. The part of the hand between the wrist and the fingers.

Mĕt-a-tar'sus. The flat of the foot.

Mī'tral. Like a miter, or bishop's cap.

Mū'coŭs mĕm'brāne. A membrane lining all the cavities of the body that are connected with the outer world.

Mū'cus. The fluid which comes from the surface of the mucous membrane.

Nar-cŏt'ic. That which benumbs and stupefies.

Nā'sal. Pertaining to the nose.

Nau'se-a. Sickness at the stomach

Nū'cle-ŭs. In anatomy, a cell within a cell.

Œ-sŏph'a-gŭs. The gullet.

Orb'it. The bony cavity in which the eye is situated.

Os in-nŏm-i-na'tum (nameless bone). The hip-bone.

Păl-pi-tā'tion. A hard, rapid beating of the heart.

Păn'cre-as. An organ of digestion. The sweet-bread in calves.

Pa-pĭl'la. A conelike prominence of the skin or mucous membrane.

Păr'a-sīte. A plant or animal that grows or lives on another.

Pa-rŏt'id. The name of a large gland under the ear.

Pa-tĕl'la. The knee-pan.

Pĕc'to-ral (Lat., *pectus*, the breast). Pertaining to the breast.

Pĕl'vis (Lat., a basin). The cavity inclosed by the hip-bones and the lower end of the backbone.

Pĕp'sin. The active principle of the gastric juice.

Pĕr-i-car'di-ŭm. A membranous bag inclosing the heart.

Pĕr-i-ŏs'te-ŭm. The membrane which covers bones.

Phā'lanx. One of the small bones of the fingers and toes.

Phăr'ynx. The throat.

Phȳs-i-ŏl'o-gy. The science of the functions of organized bodies.

Plăs'ma. The watery part of the blood.

Pleū'ra. A sac which covers the lung.

Pons Va-ro'lii (bridge of Varolius). A division of the brain which connects the other main divisions.

Pōre. The outlet of a sweat-duct.

Prŏc'ess. A bony projection from a bone

Pŭl'mo-na-ry (Lat., *pulmo*, a lung). Pertaining to the lungs.

Pū'pil. The central opening in the iris.

Pȳ-lō'rus. A muscular ring which surrounds the outlet of the stomach.

Rā'di-al. Pertaining to the radius.

Rā'di-ŭs. The outer bone of the fore-arm.

Rĕs-pi-rā'tion. The process by which oxygen is introduced into the blood, and carbonic-acid gas and vapor, and other matters, are discharged from it.

Rĕt'i-na (Lat., *rete*, a net). The terminal fibers of the optic nerve lining the back part of the eye.

Sā'crum. A part of the backbone.

Sa-lī'va. Spittle.

Sar-tō'ri-ŭs. A muscle extending from the hip to the leg, on the front of the thigh.

Scăp'ū-la. The shoulder-blade.

Scle-rŏt'ic (Lat., *scleroticus*, hard, firm). A term applied to the outer coat of the eye.

Se-bā'ceoŭs (Lat., *sebaceus*, tallowy). Applied to glands in the skin that produce a fatty fluid.

Sĕm-i-lū'nar. Shaped like a half-moon.

Sew'age. The contents of sewers.

Skĕl'e-ton. The framework of an organized body, of bone or other firm material.

Skŭll. The bony frame of the head.

Spī'nal ca-năl'. The canal in the center of the backbone.

Spī'nal cŏl'umn. The backbone.

Spī'nal cord. A cord of nerve-matter in the spinal canal.

Sprāin. An injury to the ligaments or tendons about a joint.

Stā-pē'di-us. A very small muscle in the drum of the ear.

Ster'num. The breast-bone.

Stĭm'ū-lant (Lat., *stimulus*, a goad). That which goads or excites.

Sŭb-clā'vi-an. Beneath the clavicle.

Sŭb-lĭn'gual (Lat., *sub lingua*, under the tongue). The name of a salivary gland.

Sŭb-măx'il-la-ry (Lat., *sub maxilla*, under the jaw). The name of a salivary gland.

Syn-ō'vi-a. Joint-water.

Syn-ō'vi-al mĕm'brāne. A thin membrane which lines the joint-cavity, and gives out the joint-water.

Sys-tĕm'ic. Pertaining to the general system.

Tāpe'worm. A worm that lives in the alimentary canal.

Tar'sus. A portion of the foot between the leg and the metatarsus.

Tĕn'don. A cord of white, fibrous tissue connected with a muscle.

Tĕn'don of A-chĭl'lēs. The tendon of the gastrocnemius and soleus muscles inserted in the heel. It was fabled that this was the only part in which Achilles was vulnerable.

Thō'rax. The chest.

Tĭb'i-a. The shin-bone.

Tĭb'i-al. Pertaining to the tibia.

Trans-fū'sion (Lat., *trans*, across, *fundere*, to pour). Pouring blood from the veins of one person into those of another.

Trī-chī'na. A small worm that lives in the muscles of pigs, and of some other animals, and of men.

Trī-chī-no'sis. The disease caused by trichinæ in the body.

Trī-cŭs'pid. Three-pointed.

Tўm'pa-nŭm (a drum). The middle ear.

Ul'na. The inner bone of the forearm.

Ul'nar. Pertaining to the ulna.

Vĕn'tri-cle (Lat., *ventriculus*, the belly) A name given to several small cavities in the body.

Ver'te-bra (Lat., *vertere*, to turn). One of the bones which make the backbone.

Vĭl'lus. A hairlike projection from the lining of the intestine.

Vĭt're-oŭs (Lat., *vitreus*, glassy). A term applied to the semi-fluid contents of the posterior chamber of the eyeball.

Vō'cal cords. Two fibrous bands that form the margins of the glottis.

INDEX.

INDEX.

www.ingramcontent.com/pod-product-compliance
Lightning Source LLC
Chambersburg PA
CBHW021708210326
41599CB00013B/1576
* 9 7 8 3 3 3 7 3 7 1 5 7 9 *